D0140577

SAS® Companion
for P. V. Rao's
Statistical Research Methods
in the Life Sciences

Mary Sue Younger

University of Tennessee, Knoxville

Duxbury Press

An Imprint of Brooks/Cole Publishing Company

I(T)P® An International Thomson Publishing Company

Pacific Grove • Albany • Belmont • Bonn • Boston • Cincinnati • Detroit • Johannesburg • London
Madrid • Melbourne • Mexico City • New York • Paris • Singapore • Tokyo • Toronto • Washington

Sponsoring Editor: *Cynthia Mazow*
Editorial Assistant: *Rita Jaramillo*
Production: *Dorothy Bell*

Cover Design: *Harry Voigt*
Cover Illustration: *Spike Walker/Tony Stone Images*
Printing and Binding: *Edwards Brothers, Ann Arbor*

For more information, contact Duxbury Press at 10 Davis Drive, Belmont, CA 94002, or electronically at http://www.thomson.com/duxbury.html

BROOKS/COLE PUBLISHING COMPANY
511 Forest Lodge Road
Pacific Grove, CA 93950
USA

International Thomson Editores
Seneca 53
Col. Polanco
11560 México, D. F., México

International Thomson Publishing Europe
Berkshire House 168-173
High Holborn
London WC1V 7AA
England

International Thomson Publishing GmbH
Königswinterer Strasse 418
53227 Bonn
Germany

Thomas Nelson Australia
102 Dodds Street
South Melbourne, 3205
Victoria, Australia

International Thomson Publishing Asia
221 Henderson Road
#05-10 Henderson Building
Singapore 0315

Nelson Canada
1120 Birchmount Road
Scarborough, Ontario
Canada M1K 5G4

International Thomson Publishing Japan
Hirakawacho Kyowa Building, 3F
2-2-1 Hirakawacho
Chiyoda-ku, Tokyo 102
Japan

Printed in the United States of America
10 9 8 7 6 5 4 3 2 1
ISBN 0-534-93142-1

Contents

Preface

Some History

Several years ago, I was asked to serve as a reviewer for P.V. Rao's manuscript, *Statistical Research Methods in the Life Sciences*. It became apparent to me that because Rao was trying to write a very complete and useful reference text, it was becoming large and unwieldy. In particular, any modern text purporting to instruct students on how to actually perform statistical analyses must contain reference to computer software. However, including instructions for using such software would only serve to make an already too large book even larger. Since I teach graduate students in the sciences, myself, and since I liked Rao's organization and approach, and since I know a little about SAS ® statistical programming, I volunteered to write this companion text.

How to Use this Book

What this book does, mostly, is take the examples found in P.V. Rao's *Statistical Research Methods in the Life Sciences* and show how to work them using SAS software, and interpret the results. This book should not be expected to stand on its own, but rather to be a supplement to Rao's text for those who want to use the SAS system for analyzing their data.

In most chapters, there is pretty much a one-to-one correspondence between Rao's examples and the ones in this *Companion*. Examples here are numbered with the same numbers as in Rao's text. The earlier chapters depart somewhat from this format because Rao's early chapters are more introductory or theoretical, instead of involving data analysis. In this text, Chapter 1 presents a general overview of how to write and run SAS programs, with emphasis on techniques of data entry and data set manipulations. Chapter 2 is mostly concerned with using SAS's built-in statistical functions, and Chapter 3 introduces the basic procedures for univariate data summarization. Subsequently, this text follows Rao's organization. Data management techniques not mentioned in Chapter 1 and SAS procedures not mentioned in Chapters 2 or 3 are introduced on a "just in time" basis in the later chapters.

The SAS procedures mentioned here are only a tiny fraction of SAS's collection of procedures, but they are the ones which are the "workhorses" of data analysis in the sciences. We hope that Rao's text and this *Companion* will serve as effective tools for teaching these basic techniques and as useful references in performing research and analyzing results.

It takes forever, it seems, to write and publish a manuscript. Most of this one was done using release 6.10 of SAS. By the time this book is published, the current release will be 6.12. SAS is always striving to improve and to incorporate new features, so it may well be that computations which were not available in earlier releases have been added in the current release. I have incorporated those of which I am aware.

Acknowledgments

Whenever I can't figure out how to do something in SAS, I ask Paul Wright, of the Statistics Department at the University of Tennessee, Knoxville. The program for computing Walsh averages in Chapter 5 is mostly his. Also, I think that Paul is the inventor of the terminology about "long" versus "wide" versions of data sets presented in Chapter 16.

Dr. Ralph G. O'Brien, of the Cleveland Clinic Foundation (formerly of the University of Tennessee, where I am, and also formerly of the

University of Florida, where Professor Rao is) also taught me a lot of SAS programming tricks. The method of reconstructing a data set from its sufficient statistics presented in Chapters 9 and 13 is his. Ralph also kindly provided the program for power computation shown in Chapter 7. He has expanded the power analysis program beyond what is presented here, but we did not include the expanded version because additional power issues were (in the interest of space) omitted from Rao's text.

Several reviewers have made helpful suggestions about this *Companion.* They are Mack C. Shelley, II (Iowa State University), Deborah Rumsey-Johnson (Kansas State University), Carol A. Gotway Crawford (University of Nebraska-Lincoln), and Rita Jaramillo (Wadsworth, Inc.) I would like to especially thank Ann Lehman, of the SAS Institute, Inc., for her careful analysis of the first several chapters. Everyone's kind words were much appreciated.

Mary Ann Reinstedt did a wonderful job of copy-editing the manuscript, and Cynthia Mazow has been a very patient Editor, suffering through my temper fits at the frustrations of trying to produce camera-ready copy. Finally, Alex Kugushev, former editor at Duxbury, and Prof. P.V. Rao have given me the opportunity to participate in the project. I am looking forward to meeting P.V. Rao someday!

MSY

1

Writing SAS Programs

1.1 Introduction

We begin by describing the basic structure of a SAS program, which consists of data and instructions for analysis. This is followed by an overview of how the data and instructions are submitted to the SAS system, and what output is produced. Then some basic techniques for data entry and some programming techniques for reading data are illustrated. Finally, techniques for working with SAS data sets are demonstrated.

1.2 A SAS Program

In Figure 1.1 is a simple SAS program to perform a familiar task: taking the grades of each of five students on three tests and (1) computing the class average across students for each test, and (2) computing the course average across the three tests for each student. As with any SAS program, there are four parts to Figure 1.1: (1) *optional statements* (lines 1-2), which control the layout of the output to conform to paper size, labeling of results, and so on; (2) a *data step* (lines 3-6), which prepares SAS to receive and read data, (3) the *data* itself (lines 7-11), and (4) *instructions for analysis of the data* (lines 14-19).

First, note that while in English, we begin sentences with capital letters and end them with periods, in SAS the convention is that you can use either capital or lower-case letters or a combination of them as you choose,

Figure 1.1. A SAS program to compute grades

```
options linesize=70 pagesize=54 pageno=1;
*Grades in Morning Section of Statistics;
data grades;
   input student$ test1 test2 test3;
   stuavg = (test1 + test2 + test3)/3;
lines;
ABC 100   94   90
DEF   85   88   92
GHI   95   87   89
JKL   90 100   83
MNO   79   87   88
;
run;
proc print data=grades;
run;
proc means data=grades;
   var test1 - test3;
   title 'Class Averages on Three Tests';
run;
```

and that *sentences end with semicolons*. Additionally, while it is conventional in English to indent the leading sentence of a paragraph, no such convention exists in SAS: it is a matter of style how you want to arrange your lines on the screen or page. Similarly, while in English we typically leave only one space between words, the number of spaces usually carries no information or instructions in the SAS language; SAS will read from semicolon to semicolon without regard to spacing or indentation. However, just as words must be separated by *at least one space* to be distinguishable in English, so must they be in SAS. Also, it is a matter of style whether you wish to leave spaces around signs indicating arithmetic operators, such as "+" or "=". Figure 1.1 contains examples of both styles, in line 1 (where no spaces are left around the equals sign) and line 5 (where spaces are left on either side of the equals sign).

Now let us turn to the OPTIONS statement of the program in Figure 1.1. The line

```
options linesize=70 pagesize=54 pageno=1;
```

contains the optional formatting information in this program. In line 1, the word OPTIONS alerts SAS that instructions for arranging output follow. The option LINESIZE determines how many characters can fit on a single line of output. In order for the output to fit on this page, we are restricted to 70 characters per line. For standard 8 1/2 x 11 inch paper, LINESIZE=72 works nicely. The word LINESIZE can be abbreviated as LS (or ls). PAGESIZE controls the number of lines per page: PAGESIZE=54 works well for this size paper and also for 8 1/2 x 11 inches. PAGESIZE can be abbreviated as PS (or ps).

The reason you probably want to specify line size is that, left to its own devices (the *default* condition), SAS assumes you will print your results on "greenbar", the oversized green and white striped computer paper, in "landscape" orientation. Most modern printers, however, accept standard 8 1/2 x 11 inch paper. Similarly, if you do not specify the page size, then SAS assumes that you want your printout to correspond to what shows on a single screen of your computer, and this can result in pages of output being fragmented.

More often than not, you will have to run a program more than once to get it to work just the way you want it to. SAS numbers all output pages in a single session consecutively, so that if you have a program that produces four pages of output and you submit it three times, the pages of the third set of output will be numbered 9-12. If you would rather have the pages numbered 1-4, regardless of how many times you submit the program, then use the PAGENO=1 option, which instructs SAS to reset the page number to 1 on the output whenever the program is submitted. Other options which you may want to use will be discussed later.

Another optional line in the program is

```
*Grades in Morning Section of Statistics;
```

which is a *comment*. A comment is simply a note written by the programmer to remind himself or herself about some aspect of the program. It is ignored by SAS. Comments can be written in one of two ways: (1)

beginning with an asterisk and ending with a semicolon, as shown above, or (2) beginning with /* and ending with */, as in

```
/*Grades in Morning Section of Statistics*/.
```

Comments can be written anywhere in the program, not just after the OPTIONS statement.

Not all SAS programs have the options section or comments, but all SAS programs contain at least one *data step* (or refer to a permanently stored SAS data set, but this is a more advanced technique).   However, before turning to the data step, let us first look at the data, itself. The data to be analyzed is given in these lines of the program:

```
ABC 100   94   90
DEF  85   88   92
GHI  95   87   89
JKL  90  100   83
MNO  79   87   88
```

Each line of the data is referred to as a *case* or *record* or *observation*. In this example, each case is a student, and all the information about each student is on the same line. *Note that data lines **do not** end in semicolons.* There are four pieces of information about each student, entered in the *same order* on each data line: the student's name (initials), his/her grade on Test 1, the grade on Test 2, and the grade on Test 3, respectively. These pieces of information are called *variables*.   Variables are called *numerical* or *character* according to whether their values are numbers or contain letters, respectively. In Figure 1.1, the student name is a character variable, and the three test grades are numerical variables.

It is conventional in English to *left justify* character values--that is, to line them up vertically so that they all start in the same column--and to *right justify* numerical values--line them up vertically so that the implied decimal places are all in the same column--and this is done for ease of reading in Figure 1.1. However, as long as the values are separated by at least one space, so that SAS can tell that there are four pieces of information in each row, it is not necessary to right- or left-justify values in the data set.

In case there are any *missing values* in the data set--if some student did not take one of the tests, for example--they should be indicated by a period or decimal point, by itself.

Although individual data lines do not end in semicolons, you may follow the entire data set with a semicolon, *provided it is on a line by itself*, as in Figure 1.1. It is not necessary to use this semicolon, but some programmers like it because it clearly sets off the data from the instructions that follow.

The *data step* tells SAS how to read the data. A data step begins with the word DATA. The data step in the program in Figure 1.1 consists of the following lines:

```
data grades;
   input student$ test1 test2 test3;
   stuavg = (test1 + test2 + test3)/3;
lines;
```

Since SAS programs may contain more than one data set, it is a good idea to give the different data sets different names, and this is done in the DATA statement, as in

```
data grades;
```

This line instructs SAS to construct a *SAS data set* named WORK.GRADES. It is not required for you to supply names for the SAS data sets in your program: if you write just

```
data;
```

then SAS will automatically name the data sets WORK.DATA1, WORK.DATA2, etc.

The DATA statement alerts SAS that it is to read (and create) data. The INPUT statement tells SAS how to *read* each line of the data. In Figure 1.1, the INPUT statement says that on each line of the data set you will find the name of the student, followed by that student's grades on Test 1, Test 2, and Test 3, in that order. These four variables are given names. Variable names may be one to eight characters long, can be a mixture of letters and numbers but cannot start with a number, and cannot contain spaces or symbols such as #. In Figure 1.1, the four variables are named STUDENT, TEST1, TEST2, and TEST3, respectively. SAS expects to see numerical data, unless told otherwise by following the name of the character variable with a dollar sign ($) in the INPUT statement. In Figure 1.1, there is no space left between the variable name, STUDENT, and the dollar sign, but one or more spaces could have been inserted between the name and the symbol. Note that we could have written

```
input student$ test1 - test3;
```

where TEST1-TEST3 is taken to include all values of Test$i$ where i is between 1 and 3, inclusive.

There is one more piece of information we want for each student, namely, his/her average across the three tests. Instead of computing this information by hand and typing it in, we want the computer to do this for us. ***To create new information about a case from existing information about the case, define a new variable in the data step.*** The line

```
stuavg = (test1 + test2 + test3)/3;
```

instructs SAS to create a new variable called STUAVG (student average) by adding together the grades on the three tests and dividing by 3. Note that since we did not compute and type in these values in the data set, the variable STUAVG is *not* listed in the INPUT statement-it is not input. The line above uses two arithmetic operators, addition (+) and division (/). Other arithmetic operators are subtraction (−), multiplication (*) and expo-nentiation (**). Rules of algebra apply. For example, had the parentheses been omitted, then one-third of the Test 3 grade would have been added to the total of the grades on Tests 1 and 2.

SAS has built-in functions for computing commonly used quantities such as the sum and average. The statement

```
stuavg = (test1 + test2 + test3)/3;
```

can also be written as

```
stuavg = sum(of test1 - test3)/3;
```

or as

```
stuavg = mean(of test1 - test3);
```

Other built-in functions will be mentioned throughout as appropriate. Refer to the *SAS Language: Reference* manual for a complete listing.

Note that if any of the variables TEST1-TEST3 has missing values, then SAS will have to return a missing value for STUAVG, as well.

In the program in Figure 1.1, the INPUT line and the STUAVG = ... line were both indented. This is strictly a matter of style, to make the data step more easily readable for the programmer, and carries no instructions to the computer. Similarly, the INPUT statement could have been written on the same line as the DATA statement--as long as the statements are separated by semicolons they are read as separate statements by SAS.

The data step ends with the line

```
lines;
```

which is short for "data lines follow". Synonyms for LINES are DATALINES, which is almost twice as many characters, and CARDS, which is an anachronism left over from the days when we used to compute by punching holes in cardboard.

After the data step and the data comes the word RUN. When SAS encounters this word, it executes all data steps and programming commands since the last appearance of RUN. Usually, you only really need to have RUN as the very last line of your SAS program, but it is a good habit to sort of use it like a paragraph marker in your program, to indicate the end of a related group of programming steps.

Now that SAS has been given data and instructions on how to read it and how to create new data for each case from existing data, we might want to *summarize information for each variable across cases*. This is done in SAS *procedures*, which are abbreviated "PROC". There are scores of SAS procedures, and we will study several of the available procedures for statistical analysis during the course of this text. For now, we will look at just the PRINT and the MEANS procedures.

The lines

```
proc print data=grades;
run;
```

simply request a printout of the data set. This is all that you need to write to produce the output shown in Figure 1.2. SAS takes care of formatting the arrangement of the printout on the page automatically (subject to the constraints you imposed in the OPTIONS statement). Note that the data step does not produce output. To see a printout of your data set, you have to request it using the PRINT procedure. This is because often data sets are quite large and/or you already have a printed copy of it and don't need another.

Figure 1.2. Output from the PRINT procedure

```
                         The SAS System                         1
                                    08:02 Monday, March 27, 1997

       OBS    STUDENT    TEST1    TEST2    TEST3    STUAVG

        1       ABC       100       94       90     94.6667
        2       DEF        85       88       92     88.3333
        3       GHI        95       87       89     90.3333
        4       JKL        90      100       83     91.0000
        5       MNO        79       87       88     84.6667
```

In Figure 1.2 we see the five data lines printed and arranged neatly on the page. The four pieces of information (STUDENT, TEST1, TEST2, TEST3) that we entered for each case, and also the information on STUAVG which we had SAS compute for us, are given for each student. By default, data for all variables and for all cases will be printed. To specify only selected variables, list them in the VARiables statement. For example,

```
proc print;
   var student stuavg;
run;
```

will result in only the students' names and course averages being printed. Note also that you don't need to follow the variable name STUDENT with a $, since once you have designated a variable as a character variable in the INPUT statement, you don't need to do it again. See Section 1.6 for how to print only a subset of the cases.

The output in Figure 1.2 has the default heading, "The SAS System" and the page number 1 (because the option PAGENO=1 was used) in the upper right-hand corner. The date and time that the program was run are shown next. (Date and time information can be suppressed using the OPTIONS NODATE statement, and they will be omitted in subsequent figures in this text.) Provided the output file was saved as a "print file" (how this is done depends on the hardware and software you are using), Figure 1.2 will be all that is printed on a single sheet of paper.

Unless otherwise instructed, SAS assumes that any procedure is to be executed on the *most recently created data set*. Therefore, since there is only one data set in the program in Figure 1.1, we could have omitted the DATA=GRADES from the PROC PRINT command. However, it is good programming practice to get into the habit of specifying the data set to be used, because sometimes it is easy to forget that you have multiple data sets in a program. Note also that although SAS names the data set data set WORK.GRADES, it can be referred to as just GRADES in the DATA = statement.

Finally, the lines

```
proc means data=grades;
   var test1 - test3;
   title 'Class Averages on Three Tests';
run;
```

produce the output in Figure 1.3. The MEANS procedure can be used to obtain a variety of descriptive statistics for the variables in the data set. What is given to you automatically (called the *default* output) is determined by the LINESIZE option. Using LS=70, we get, for each of the three tests, the number of students, mean, standard deviation, and minimum and maximum values. Other statistics could have been requested as well, as we will see later (in Chapter 4, for example). The MEANS procedure will give statistics for *all numerical* variables in the data set. (Obviously, it is impossible to compute the mean student name!) In this case, however, we did not want to include the *final* grades of the students when computing the average, etc., so we specified in the VARiables statement that we wanted the procedure performed on just TEST1, TEST2, and TEST3.

Figure 1.3. Output from the MEANS procedure

```
                    Class Averages on Three Tests                  2

Variable  N         Mean        Std Dev        Minimum        Maximum
-------------------------------------------------------------------
TEST1     5    89.8000000      8.2280010     79.0000000    100.0000000
TEST2     5    91.2000000      5.7183914     87.0000000    100.0000000
TEST3     5    88.4000000      3.3615473     83.0000000     92.0000000
-------------------------------------------------------------------
```

A TITLE statement was also used to label page 2 of the printout. Simply write the word TITLE and then enclose the desired title in single quotes (and don't forget the semicolon after the close quotes). If the title is a long one, break it up into separate lines, designated TITLE1, TITLE2, etc. A title will stay in effect and will label all subsequent pages until it is overridden by a new title. The easiest way to simply turn off all titles is to write

```
title ' ';
```

or just

```
title;
```

in the PROC step in which you want this to take effect.

1.3 Running a SAS Program

The SAS System is available for a wide variety of computers, from mainframes to laptops, and exactly what commands you issue to submit programs, save output, print output, etc. vary according to the kind of hardware and software you are using. Whatever the environment, however, SAS works in basically the same way.

First, a *program file* is created by the programmer, using some text editor. Figure 1.1 is an example of a program file--it contains data and instructions for analysis of the data. (Note: Sometimes it is convenient to have the data and the instructions for analysis in separate files. This will be discussed in the next section.) The program file is submitted to the SAS System for execution (exactly how this is done varies according to whether you are working on a mainframe computer, in DOS or Windows, etc.). The computer processes the data according to the instructions in the program file and puts the results in an *output* file. Figures 1.2 and 1.3 show the two pages of the output file that was created when the program in Figure 1.1 was executed. Additionally, the computer creates another file, called the *log* file, that lists the program statements and writes comments concerning both the execution of the DATA step and PROC steps, and the discovery of any programming errors. Figure 1.4 shows the log file produced by execution of the program in Figure 1.1. How to look at, save, and print these three files varies somewhat according to the equipment you are using.

Let us take a look at the information provided in the log file. The first three lines in Figure 1.4 give information about the SAS copyright and the release number and licensing of the copy of the software you are running. The lines of the program file are numbered and reproduced in the log file. Note that the log file does not print the data lines of the program file, although it has assigned numbers to the data lines. If you have a relatively small data set, you might want it to be printed in the log file. To do this, write the command

```
list;
```

Figure 1.4. SAS log file for the program in Figure 1.1.

```
NOTE: Copyright(c) 1989 by SAS Institute Inc., Cary, NC USA.
NOTE: SAS (r) Proprietary Software Release 6.08   TS404
      Licensed to UNIVERSITY OF TENNESSEE, Site 0009122001.


1     options linesize=70 pagesize=54 pageno=1;
2     *Grades in Morning Section of Statistics;
3     data grades;
4       input student$ test1 test2 test3;
5       stuavg = (test1 + test2 + test3)/3;
6     lines;

NOTE: The data set WORK.GRADES has 5 observations and 5 variables.
NOTE: The DATA statement used 8.06 seconds.


12    ;
13    run;
14    proc print data=grades;
15    run;

NOTE: The PROCEDURE PRINT used 5.0 seconds.


16    proc means data=grades;
17      var test1 - test3;
18      title 'Class Averages on Three Tests';
19    run;

NOTE: The PROCEDURE MEANS used 3.97 seconds.
```

in the data step. After the data step (lines 3-6), the log file makes a note of how many cases and how many variables it understood. In this case, the WORK.GRADES data set had five observations (students) and five variables. The five variables are the four that were read by the INPUT statement, namely STUDENT, TEST1, TEST2, and TEST3, and also the one that SAS was instructed to create, STUAVG. It is a good habit to check the log file to make sure that no data lines were inadvertently omitted and that SAS understood the instructions for creating new variables.

Subsequent portions of the log file are divided according to portions of the program that begin with PROC and end with RUN;. After each portion of the program, SAS prints a message as to how long it took to execute the commands.

Should your program contain any errors, the log file is where you look for help in discovering where the errors are. However, given the huge number of errors that programmers can make, it is not always easy for SAS to either figure out what went wrong or to tell you where the errors are. Error messages can be quite cryptic to the novice user--and sometimes even to experienced users! But experience does help in figuring out what the error messages are trying to say. To get you started, let us purposely make some errors in the program in Figure 1.1 and see what happens in the log file.

Figure 1.5 is a program with some common errors. For instance, note that the dollar sign has been omitted after STUDENT in the INPUT statement. Recall that by default SAS will then expect to find *numerical* values for the variable, STUDENT.

The most common error made by beginner SAS programmers is to forget the semicolon at the end of a SAS "sentence". (It is amazing, however, how quickly one gets into the habit of ending SAS statements with semicolons, and even more amazing that you do not transfer this habit to writing English compositions!) In Figure 1.5, a comment line has been added in the data step reminding us that the variable STUAVG is the average of the three test grades for each student; however, note that the novice programmer forgot to end this comment with a semicolon.

In the PRINT procedure in Figure 1.5, notice that PROC was misspelled as PORC and that the semicolon was omitted from that line. Another common error is to forget to close quotes, as was done in the TITLE statement in the MEANS procedure. Look at the log file for this program, as shown in Figure 1.6.

The first problem detected in the log file is that there is invalid data for the variable STUDENT. Since STUDENT was not followed by a dollar sign, SAS expected its values to be numerical, rather than character. The log file says that there is invalid data for this variable in line 8 (the first data line), columns 1-3. It then gives you a ruler to help you count columns. The plus sign is at column 5, the symbol 1 indicates column 10, the next plus

Figure 1.5. A SAS program with errors

```
options linesize=70 pagesize=54 pageno=1;
*Grades in Morning Section of Statistics;
data grades;
   input student test1 test2 test3;
   *stuavg is the average of the three test
      grades for each student
   stuavg = (test1 + test2 + test3)/3;
   lines;
ABC 100   94   90
DEF  85   88   92
GHI  95   87   89
JKL  90 100   83
MNO  79   87   88
;
run;
porc print data=grades
run;
proc means data=grades;
   var test1 - test3;
   title 'Class Averages on Three Tests;
run;
```

sign is at column 15, the 2 indicates column 20, and so on. Then the data line
is reproduced as entered in the program. Following that, SAS says that the
value for STUDENT has been converted to a missing value (indicated by a
period) TEST1 has the value 100, TEST2 has the value 94, and TEST3 has
the value 90.

ERROR and _N_ are *automatic variables.* These are variables
created automatically by the data step. To oversimplify a bit, _ERROR_
counts the number of errors in a data line and _N_ numbers the data lines.
SAS encounters the same problem--expecting a numerical value and finding
a character value--on the next data line, so it goes through the whole process
of giving the invalid data message and reproducing the data line again, and
continues to do so for each subsequent data line.

Figure 1.6. Log file showing errors

```
1      options linesize=70 pagesize=54 pageno=1;
2      *Grades in Morning Section of Statistics;
3      data grades;
4        input student test1 test2 test3;
5        *stuavg is the average of the three test grades for each
                student
6.       stuavg = (test1 + test2 + test3)/3;
7        lines;

NOTE: Invalid data for STUDENT in line 8 1-3.
RULE:----+----1----+----2----+----3----+----4----+----5----+----6----+
8     ABC 100   94   90
STUDENT=. TEST1=100 TEST2=94 TEST3=90 _ERROR_=1 _N_=1
NOTE: Invalid data for STUDENT in line 9 1-3.
9     DEF   85   88   92
STUDENT=. TEST1=85 TEST2=88 TEST3=92 _ERROR_=1 _N_=2
NOTE: Invalid data for STUDENT in line 10 1-3.
10    GHI   95   87   89
STUDENT=. TEST1=95 TEST2=87 TEST3=89 _ERROR_=1 _N_=3
NOTE: Invalid data for STUDENT in line 11 1-3.
11    JKL   90 100   83
STUDENT=. TEST1=90 TEST2=100 TEST3=83 _ERROR_=1 _N_=4
NOTE: Invalid data for STUDENT in line 12 1-3.
12    MNO   79   87   88
STUDENT=. TEST1=79 TEST2=87 TEST3=88 _ERROR_=1 _N_=5
NOTE: The data set WORK.GRADES has 5 observations and 4 variables.
NOTE: The DATA statement used 11.06 seconds.



13    ;
14    run;
15    porc print data=grades
      ----
      14
16    run;
      ---
      202

WARNING 14-169: Assuming the symbol PROC was misspelled as PORC.

ERROR 202-322: The option or parameter is not recognized.

NOTE: The SAS System stopped processing this step because of errors.
NOTE: The PROCEDURE PRINT used 3.02 seconds.



17    proc means data=grades;
18      var test1 - test3;
19      title 'Class Averages on Three Tests;
20    run;
```

The next error is not recognized by SAS as being an error. Since the semicolon was left off the comment about STUAVG, SAS took it that the comment was

```
*stuavg is the average of the three test grades for
each student
stuavg = (test1 + test2 + test3)/3;
```

since it read from the asterisk to the next semicolon. Thus, it never recognized the instructions to compute values for the variable STUAVG. The programmer would know that something was wrong upon reading that the data set had four variables, instead of five.

In line 15, the misspelled word PORC is underlined and given a *warning number* 14. Catalogs of warning numbers and error numbers exist, but recognizing that you are not likely to be carrying these around with you, after line 16 SAS provided the following interpretation of a Warning 14:

```
Assuming the symbol PROC was misspelled as PORC.
```

This must be a pretty common error, so SAS proceeds under the assumption of a misspelling and continues to try to run the program. However, immediately another problem is encountered, so that the word RUN is underlined in line 16 and labeled as an Error 202. Looking down, we see that errors numbered 202-322 have to do with

```
The option or parameter is not recognized.
```

What this means is that SAS doesn't know what the word RUN means when it is part of a PROC statement. Usually this message is a clue that you have left off a semicolon, as was indeed the case in line 15. One would think that this would be a common error, like misspelling PROC as PORC, and that SAS would be programmed to say something to the effect of, "Assuming semicolon left off previous statement," but such is not the case. At this point, SAS is so confused that it gives up:

```
NOTE: The SAS System stopped processing this step because
of errors
```

and tries to execute the next procedure.

The log file shows no error messages in the MEANS procedure, but neither is there any note about how many seconds the procedure took to execute nor was any output file created. The problem is that SAS thinks that everything between the open and close quotes is part of the title, even though it never finds the close quotes symbol. Thus, it never sees the RUN command and consequently, never runs the MEANS procedure.

You won't break your computer by purposely writing programming errors, and doing this is a good technique for learning how to interpret error messages in the log file. Even experts sometimes have trouble figuring out what is wrong with a program, so don't be discouraged, even if you are frustrated.

1.4 Using the INFILE Statement

Figure 1.1 contained both a data set and the instructions for its analysis. Provided the data set is small, as was the case in Figure 1.1, and/or the data are to be analyzed only once, it is efficient to have the data and the program all in one file. However, suppose that instead of having information on just five students, the data contained test scores for *500* students in a very large lecture section. It would be very cumbersome to move around that data set, writing data step statements before it and procedure steps after it. Also, it might make you nervous to think that a typographical error would mess up this big data set. Many research data sets are quite large as well, and often many different analyses are run on them, typically at different points in time. In situations like these, it is more efficient to have the data reside in a separate file from the programming statements to analyze it. An INFILE statement in the data step in a program file instructs SAS to apply the programming statements to the data in the specified data file.

Although the data set in Figure 1.1 is quite small, we can use it to illustrate how the INFILE statement works. Create a data file that contains *only* the data to be analyzed--no input statements, headings, comments, or anything else. Suppose we put the data for the five students in the morning section of the statistics course into a file called AMSTAT.DAT. This data file is shown in Figure 1.7. The program in Figure 1.8 shows how to access this data file using the INFILE statement. Comparing this program to the one in Figure 1.1, we note the following differences of any consequence: First, the INFILE statement follows the DATA statement. The data file is referred to in single quotes. There is no CARDS or LINES statement. And finally, there are no data lines. The program in Figure 1.8 will produce exactly the same output as in Figures 1.2 and 1.3.

Figure 1.7. The AMSTAT.DAT data file

```
ABC 100   94   90
DEF  85   88   92
GHI  95   87   89
JKL  90  100   83
MNO  79   87   88
```

Figure 1.8. Using the INFILE statement

```
options ls=70 ps=54 pageno=1;
*Grades in the Morning Section of Statistics;
data grades;
   infile 'amstat.dat';
   input student$ test1 - test3;
   stuavg = (test1+test2+test3)/3;
run;
proc print data=grades;
run;
proc means data=grades;
   var test1 - test3;
   title 'Class Averages on Three Tests';
run;
```

At this point, it might be well to point out the distinction between a *data file* and a *SAS data set*, since this is sometimes a source of confusion for beginning users. The file AMSTAT.DAT is a *non-SAS* file. It was created by a text editor and stored. SAS procedures can process only data stored in SAS data files. SAS files, which include output and log files as well as data files, can only be created by SAS. We have already seen how SAS created output and log files. SAS data files are created by the DATA step in a SAS program. Using either *in-stream data*, as in Figure 1.1, or data stored in a separate data file, the DATA step creates a *SAS data set* called WORK.GRADES. SAS procedures are then executed on the SAS data set.

Sometimes, SAS procedures themselves can be used to create new SAS data sets, which contain statistics computed by the procedures. An example of this will be shown in Section 1.6.

1.5. Data Entry Techniques

Once you catch on about SAS programming statements, you will find that SAS programs are relatively easy to write. This is one reason that SAS has become such a large, multipurpose system. It is often the case that the task of typing in the data to be analyzed is several times larger than writing the instructions for analysis. In this section, we will show a few techniques for entering data and writing input statements to read it.

Multiple Cases per Line

The most straightforward way to enter data is described in Section 1.2. Each line in a data file contains information on a single case, or record, subject, or observation. The pieces of information (variables) about each case appear in the same order on each data line. This straightforward method is often quite inefficient, however.

For example, suppose that you have test grades of 20 students on a single test, and you simply want to compute the mean and standard deviation of the grades. Which student earned what grade is unimportant. According to the one-case-per-line rule, the data step and the data would look like this:

```
data test1;
```

```
   input grade;
   lines;
87
98
78
100
 .  .  .
88
;
proc means data=test1;
  var grade;
run;
```

There are 20 data lines (not all shown), one for each of the 20 students. A more efficient way to enter the data would be as *multiple cases per line*,

```
87 98 78 100 . . . 88
```

which would take only one or two lines. The question is, how can you write an input statement to read data like this? The symbol

```
@@
```

in an input statement instructs SAS to stay on the same line and continue reading values of the variable(s) preceding the *double trailing @*. Thus, we would write

```
data test1;
  input grade @@;
lines;
```

This technique can also be employed when there are several pieces of information on each case. For example, the student names and the three test grades in the AMSTAT.DAT data set could have been entered as

```
ABC 100 94 90 DEF 85 88 92 GHI 95 87 89 JKL 90 100 83
MNO 79 87 88
```

if the INPUT statement had read

```
input student$ test1 - test3 @@;
```

This INPUT statement instructs SAS to read the values of the four variables, stay on the same line and read another four values, and so on.

Multiple Lines per Case

At the other end of the spectrum from having multiple cases on a single data line is the instance of having so much information on each case that it takes multiple data lines to enter it all. For example, suppose that for each student, the name and social security number are entered on one data line, and then grades on each of three tests are entered on the next data line. The slash in the statement

```
input student$ ssn$/test1 - test3;
```

tells SAS to read the values of the first two variables and then go to the next data line to read the next three values for that case.

Data in Subgroups

Suppose there are four sections of a lab course, each containing six students. The grades earned by the students were:

Section A: 76 90 89 100 92 83
Section B: 100 77 88 90 88 85
Section C: 90 92 85 87 90 100
Section D: 87 94 73 95 85 90.

One way to enter data on both the student's grade and his/her section is the one-case-per line method:

```
data;
  input section$ grade;
lines;
```

```
A 76
A 90
A 89
 . . .
D 85
D 90
;
```

This is too tedious, and the following is not much of an improvement:

```
data;
   input section$ grade @@;
lines;
A 76 A 90 A 89 . . . D 85 D 90
;
```

What the researcher would like to do is input the data essentially as given:

```
A   76   90   89 100   92   83
B  100   77   88   90   88   85
C   90   92   85   87   90  100
D   87   94   73   95   85   90
```

The question is, how can you write an input statement to read these data?

Consider the following:

```
data;
   input section$ @@;
      do i = 1 to 6;
         input grade @@;
         output;
      end;
lines;
```

The INPUT statement instructs SAS to read the first value on each data line as the value of a character variable, SECTION, and to stay on the same data line. The lines beginning with DO and ending with END are called a *do group*. The letter *i* tells how many *iterations* of the following SAS commands to perform. (Any letter--not just *i*--can be used as the index indicating the number of iterations.) In this case, it says to input a GRADE and output the value to the data set six times before moving on to the next data line and reading another value for SECTION. (Why the OUTPUT statement is needed is somewhat technical and will not be discussed here. A very good elementary discussion of how the data step is executed can be found in *How SAS Works,* by Paul A. Herzberg.)

If the PRINT procedure is performed after the above data step, the output looks like Figure 1.9. Recall that the PRINT procedure will print one case per line, regardless of how the SAS data set is set up in the program. Note that in addition to the variables SECTION and GRADE, the variable I has been added to the data set. If you don't want or need to have a record of the iteration number, simply write

```
drop i;
```

after the END statement.

Figure 1.9. Printout of data entered using a DO statement

OBS	SECTION	I	GRADE
1	A	1	76
2	A	2	90
3	A	3	89
4	A	4	100
5	A	5	92
6	A	6	83
7	B	1	100
8	B	2	77
9	B	3	88
10	B	4	90
11	B	5	88
12	B	6	85
13	C	1	90
14	C	2	92
15	C	3	85
16	C	4	87
17	C	5	90
18	C	6	100
19	D	1	87
20	D	2	94
21	D	3	73
22	D	4	95
23	D	5	85
24	D	6	90

If the four sections all have differing numbers of students in them, then a bit more programming is needed. Suppose we add one student to Section A and two students to Section C:

```
A   76   90   89 100   92   83   91
B  100   77   88   90   88   85
C   90   92   85   87   90  100   84   89
D   87   94   73   95   85   90
```

Add an impossible "data" value at the end of each of the first three data lines, and a different impossible "data" value at the end of the last data line:

```
A   76   90   89 100   92   83   91   -1
B  100   77   88   90   88   85   -1
C   90   92   85   87   90  100   84   89   -1
D   87   94   73   95   85   90   -2
```

There is no way that a *bona fide* grade of either −1 or −2 will be encountered in the data. These impossible values are added as "flags", as you will see by examining the data step below:

```
data;
   a: input section$ @@;
   b: input grade @@;
      if grade = -1 then go to a;
      if grade = -2 then stop;
      else output;
   go to b;
lines;
```

First, note that the two INPUT lines are given *labels a* and *b* by writing the name of the label, followed by a colon, at the beginning of the lines. One-letter labels were used here for simplicity, but longer, more descriptive labels could have been used. For example, the first INPUT statement might have been labeled

```
readsec:
```

since it *reads* the *section*. Note, however, that these labels are *not* interpreted as instructions by the computer. The first INPUT line instructs

SAS to read a value of the character variable SECTION and stay on the same line. The next INPUT statement instructs SAS to then read a value of the numerical variable GRADE and stay on the same line. At this point, SAS has read

section = A, grade = 76.

Now, the IF-THEN/ELSE statements have SAS perform a check on the value it just read and take different routes depending on what the value is. The general form of IF-THEN/ELSE commands is

IF *condition* THEN *statement1;*
ELSE *statement2;*

where if *condition* is "true", then *statement1* is executed; but if *condition* is "false", then *statement2* is executed. The IF-THEN/ELSE commands in the data step above say that if the value just read is −1, then this means that the end of a data line has been reached and SAS should go to the data step line labeled *a.* This line instructs SAS to go to the next data line and read a new value for SECTION. Or, if the last value just read is a −2, this means that the end of the last data line has been reached, so SAS can stop reading data lines. Neither of these was the case, since the last data value read was a 76, so the ELSE statement says to output the 76 to the data set and then go to the data step line labeled *b,* where it reads the next value of GRADE. Then the series of checks is repeated, until all data are read. Note that the −1 and −2 never get outputted to the data set. They are simply flags that direct SAS to either go to the next data line or to stop reading data.

Another way to accomplish the same task is to use an INFILE statement with the MISSOVER option. The following program will read the data with differing numbers of students in each section:

```
data;
  infile cards missover;
  input section $ @;
    do until (grade eq .);
      input grade @;
      if grade ne . then output;
```

```
        end;
lines;
A   76   90   89 100   92   83   91
B  100   77   88   90   88   85
C   90   92   85   87   90  100   84   89
D   87   94   73   95   85   90
;
run;
```

(Note: "eq" is short for "equals" and "ne" is short for "not equals". See Figure 1.12.) Note that the INFILE statement is used, even though the data are given as part of the program, instead of residing in a separate data file. The CARDS statement is what tells SAS that the data will follow. The MISSOVER option assigns a missing value (.) when SAS runs out of data values to read--that is, when it reaches the end of a line.

The INPUT statement instructs SAS to read SECTION as a character variable and stay on the same line. The INPUT and IF statements tell SAS to then read a value for GRADE (a numerical variable) stay on the same line and read another one, and as long as there is a nonmissing value for GRADE, to output the value to the data set. What tells SAS to quit is the DO group. It tells SAS to read and output values for GRADE UNTIL it encounters a missing value for GRADE; then it will read another value for SECTION. The value in the parentheses in the UNTIL command is evaluated every time SAS reads a grade; when the value in the parentheses is "true", the DO group is not executed.

Formatted Input

In the examples of data files we have used so far, values of different variables on a given line have been separated by blank spaces. This can be a very convenient way of entering data in some cases, but a very inefficient method in others. For example, if there is a very large amount of data to be entered, it might increase the time and number of lines required for typing in the data considerably if spaces must be left between values. In such instances, data lines might look like

```
abc999490
def858892
ghi958789
jkl909983
mno798788
```

These data can still be read, *as long as the value of any variable always occupies the same columns on each line.* This is called *formatted input.* All that is required is that the input statement tells what columns to look in to find the values for each of the variables, such as in

```
input name$ 1-3 test1 4-5 test2 6-7 test3 8-9;
```

If formatted input is used, numerical values must be right justified. For example, if the computer is instructed to read a value from columns 1-3, and a 7 is in column 1, with columns 2 and 3 blank, the value will be read as 700.

Another example of when it is convenient to specify the columns in which values are to be found is when data look like:

```
Abner B. Caldwell        100 94 90
David E. Fields          85 88 92
George H. Inman          95 87 89
John K. Lawson           90 100 83
Murray N. Oppenheimer    79 87 88
```

There are four pieces of information on each case: the name, and the grades for each of three tests. However, writing an input statement

```
input name$ test1 test2 test3;
```

will result in the computer assigning the value "Abner" to the variable NAME, and then trying to assign the value "B." to the variable TEST1, "Caldwell" to TEST2 and 100 to TEST3. Provided that on each line, the first 28 spaces/columns are reserved for the students' names, then the input statement can read

```
input name$ 1-28 test1-test3;
```

Imbedded Blanks

But what if the data have already been entered as shown below?

```
Abner B. Caldwell  100 94 90
David E. Fields    85 88 92
George H. Inman    95 87 89
John K. Lawson     90 100 83
Murray N. Oppenheimer  79 87 88
```

As long as *at least two* blank spaces separate the NAME field from the field for TEST1 in the data lines, the *ampersand symbol (&)* indicates imbedded blanks in the values for the NAME variable:

```
input name $ & test1-test3;
```

Note that the & symbol follows the variable name and the $ sign to which it applies.

1.6 Working with SAS Data Sets

SAS data sets can be subsetted, combined with other SAS data sets, created from existing SAS data sets, and created by SAS procedures.

Creating One Data Set from Another

Consider the AMSTAT.DAT data set of Figure 1.7. The scenario is that these are grades of students enrolled in a morning section of a statistics course. Suppose we want to use these data to create a new data set, which has all the information in the AMSTAT.DAT data, but in addition assigns the grade of A to students whose course averages are 90 or above, Bs to students with averages in the 80s, and so on. We will call this data set REPORT. Consider the program in Figure 1.10 and the output in Figure 1.11.

The SAS data set WORK.REPORT contains all the cases and variables as does the data set WORK.AM, and in addition contains the variables STUAVG and GRADE. Of course, these new variables could have been defined in the data step that created the SAS data set WORK.AM from the AMSTAT.DAT data file, but the point of this exercise is to illustrate how a new SAS data set (REPORT) can be created from an existing SAS data set (AM).

The program in Figure 1.10 also illustrates the use of IF-THEN statements to define a new variable. (In the last section, we saw how IF-THEN/ELSE statements could be used to direct the computer to different lines in a data file.) Note that since the variable being defined, GRADE, is a

Figure 1.10. Program illustrating the creation of a new data set from an existing one

```
options ls=70 ps=54 pageno=1;
data am;
   infile 'amstat.dat';
   input student$ test1 - test3;
run;
data report;
   set am;
   stuavg = (test1+test2+test3)/3;
   if stuavg >= 90 then grade = 'A';
   if 80 <= stuavg <= 89 then grade = 'B';
   if 70 le stuavg le 79 then grade = 'C';
   if 60 le stuavg le 69 then grade = 'D';
   if stuavg < 60 then grade = 'F';
run;
proc print data=report;
run;
```

Figure 1.11. New data set created from an existing one

OBS	STUDENT	TEST1	TEST2	TEST3	STUAVG	GRADE
1	ABC	100	94	90	94.6667	A
2	DEF	85	88	92	88.3333	B
3	GHI	95	87	89	90.3333	A
4	JKL	90	100	83	91.0000	A
5	MNO	79	87	88	84.6667	B

character variable, its values are enclosed in single quotes in the IF-THEN commands.

   Comparison operators are also employed in the IF-THEN commands in Figure 1.10. Because some keyboards are limited in the symbols they provide, there are alternative ways to indicate equality or inequality. Figure 1.12 is a table of comparison and logical operators in SAS.

Figure 1.12. Comparison and logical operators

```
Description                             Symbol
Comparison Operators
   equal to                             =  or EQ
   not equal to                         ^= or NE
   greater than                         >  or GT
   less than                            <  or LT
   greater than or equal to             >= or GE
   less than or equal to                <= or LE
   equal to one of a list               IN
Logical Operators
   logical and                          &  or AND
   logical or                           |  or OR
   logical not                          ^  or NOT
```

   Related to the IF-THEN/ELSE command is the *subsetting IF* command. This is useful for creating a new data set that contains only a specified subset of the original data set. The general form of the subsetting IF is

IF expression;

If *expression* is "true", then the data step continues processing those cases; otherwise, SAS ignores the cases and does not output them to the data set. For example, to choose only those students in the morning section of the course who earned a grade of "A", write

```
data a;
  set report;
  if grade = 'A'; run;
```

Closely related to, but actually quite different from, the subsetting IF command is the WHERE statement. Its form is

WHERE *expression;*

where the expression is a valid arithmetic or logical expression that specifies the condition that must be met in order for SAS to select the observation *from a SAS data set* to be used in executing a procedure. The differences between IF and WHERE are: (1) IF can only be used in a data step, whereas WHERE can be used in a data step *or* in a PROC step, and (2) the WHERE statement selects observations from existing SAS data sets only and therefore cannot be used to *create* a new SAS data set from raw data in conjunction with an INPUT statement.

In Section 1.2 it was demonstrated how to print only a subset of the *variables* in a data set by specifying the variable names in the VAR statement in the PRINT procedure. The WHERE statement can be used to obtain a printout of a subset of the *cases* in a data set. Recall from Figure 1.6 that an *automatic variable,* _N_, which essentially counts the number of cases in a SAS data set, is automatically created whenever a data step is executed. This variable is not, however, outputted to the data set. But it can be used to define a variable in the SAS data set. For example,

```
data grades;
  input student$ test1-test3;
  n = _n_;
lines;
```

will result in the data set WORK.GRADES containing the five variables, STUDENT, TEST1, TEST2, TEST3, and N. Then

```
proc print data=grades;
  where n <=3;
run;
```

will print the data for only the first three cases.

Combining Data Sets

Suppose now that in a data file named PMSTAT.DAT there is information about grades of students in an afternoon section of the statistics course, and that the information is entered in the same way as in the AMSTAT.DAT file. The data step below shows how to combine these two data files into a *single* SAS data set.

```
data am;
  infile 'amstat.dat';
  input name$ test1-test3;
run;
data pm;
  infile 'pmstat.dat';
  input name$ test1-test3;
run;
data both;
  set am pm;
run;
```

A printout of the data set WORK.BOTH will show the cases of the WORK.PM data set appended to the end of the WORK.AM data set.

Sorting Data Sets

This time suppose that information from both sections is stored in a single file, STAT.DAT. For each student, the information available is the student's name, the section in which he/she is enrolled, and his/her grades on the three tests. What is required is a separate analysis for each of the two sections. There are several ways this could be accomplished. Separate data sets for the two sections could be created using the subsetting IF or the WHERE command, and then each data set could be analyzed separately. However, the SORT procedure can be used so that one set of commands can be used to apply to both subsets of the data file separately. Suppose the STAT.DAT data set is as shown in Figure 1.13.

Figure 1.13. Data set containing information from two sections of a statistics course

```
ABC am 100   94    90
DEF am  85   88    92
GHI pm  95   87    89
JKL am  90  100    83
MNO pm  79   87    88
```

Recall that the MEANS procedure in either Figure 1.1 or in Figure 1.8 will produce the test averages, standard deviations, etc. in Figure 1.3, each based on n = 5 students. To obtain separate analyses for the three "am" students and the two "pm" students, we first SORT the data set BY SECTION and then perform the MEANS procedure BY SECTION, as shown in Figure 1.14. Figure 1.15 shows the output.

Figure 1.14. Using the SORT procedure

```
options ls=70 ps=54 pageno=1;
data grades;
   infile 'stat.dat';
   input name$ section$ test1-test3;
run;
proc sort data=grades;
  by section;
run;
proc means data=grades;
  var test1-test3;
  by section;
run;
```

Note that the SORT procedure produces no output. Rather, it rearranges the SAS data set into subgroups according to the value of the BY variable--SECTION, in this case. Then, when a procedure is executed BY that variable, separate analyses are produced for each subgroup, as is shown in Figure 1.15. If you want to use a BY statement, the SAS data set must have observations grouped together according to the values of the BY variable.

Figure 1.15. Grades analyzed by section

```
--------------------------- SECTION=am ---------------------------


Variable   N          Mean        Std Dev       Minimum        Maximum
-----------------------------------------------------------------------
TEST1      3     91.6666667      7.6376262     85.0000000    100.0000000
TEST2      3     94.0000000      6.0000000     88.0000000    100.0000000
TEST3      3     88.3333333      4.7258156     83.0000000     92.0000000
-----------------------------------------------------------------------


--------------------------- SECTION=pm ---------------------------


Variable   N          Mean        Std Dev       Minimum        Maximum
-----------------------------------------------------------------------
TEST1      2     87.0000000     11.3137085     79.0000000     95.0000000
TEST2      2     87.0000000              0     87.0000000     87.0000000
TEST3      2     88.5000000      0.7071068     88.0000000     89.0000000
-----------------------------------------------------------------------
```

When the SORT procedure is performed, the subsets of the data set are arranged in numerical value from smallest to largest if the BY variable is a numerical variable, or in alphabetical order if the BY variable is a character variable. Figure 1.15 shows the two subsets of the GRADES data set, with the "am" section first, in alphabetical order of the values of the character variable SECTION. If you want an alphabetical listing of students within each section, first sort by section and then by name. This can be done with one SORT command:

```
proc sort;
  by section name;
```

Outputting Results to Data Sets

Often, statistics computed from raw data by some SAS procedure are used as the raw data for another procedure. For example, suppose that the test means for the two sections of the statistics course are to be graphed after they have been computed. A SAS procedure will be used to construct the graphs. But the data accessed by a SAS procedure must be stored in a

SAS data set. The question is, how can we get the statistics computed by the MEANS procedure into a SAS data set to be used by another procedure?

Many SAS procedures provide for their results to be *outputted* to a SAS data set. The command for this is OUTPUT. You can specify the name of this new data set by using OUT=*name*, where *name* is the data set name supplied by the programmer. Then the statistics to be outputted are listed and given variable names by the programmer. Figure 1.16 shows a program to compute class averages on Test 3 for the "am" and "pm" sections and output these averages under the name YBAR to a data set named WORK.AVGS. Figure 1.17 shows the output from the MEANS procedure and Figure 1.18 shows the contents of the WORK.AVGS data set.

Figure 1.16. Program for outputting computed means to a new data set

```
options ls=70 ps=54 pageno=1;
data grades;
   infile 'stat.dat';
   input name$ section$ test1-test3;
run;
proc sort data=grades;
   by section;
run;
proc means data=grades;
   var test3;
   by section;
   output out=avgs mean=ybar;
run;
proc print data=avgs;
run;
```

We see in Figure 1.18 that this data set has two observations. For each case, there is information on the SECTION (since we sorted by section), the mean (YBAR) for each section, and the number of cases (_FREQ_) on which the mean was based. (_TYPE_ = 0 is nothing of interest to us here.) The variables SECTION, _FREQ_ and YBAR are variables which can be used in any SAS procedure that refers to the data set WORK.AVGS.

Figure 1.17. Test 3 means by section

```
Analysis Variable : TEST3


---------------------------- SECTION=am ----------------------------


    N           Mean        Std Dev       Minimum       Maximum
    ---------------------------------------------------------------
    3       88.3333333     4.7258156     83.0000000    92.0000000
    ---------------------------------------------------------------


---------------------------- SECTION=pm ----------------------------


    N           Mean        Std Dev       Minimum       Maximum
    ---------------------------------------------------------------
    2       88.5000000     0.7071068     88.0000000    89.0000000
    ---------------------------------------------------------------
```

Figure 1.18 Outputted data set

OBS	SECTION	_TYPE_	_FREQ_	YBAR
1	am	0	3	88.3333
2	pm	0	2	88.5000

Merging Data Sets

Finally, consider the case in which the test grades of students in the morning section are stored in a file named AMSTAT.DAT, but the homework grades for these same students are stored in a separate data file, named AMHWK.DAT, and we want to create a single data file that has both the homework and the test grades for all of the students. This is called *match merging* the two files, and is illustrated below:

```
data tests;
   infile 'AMSTAT.DAT';
   input name$ test1-test3;
run;
data hwk;
   infile 'AMHWK.DAT';
   input name$ hwk1 - hwk 5;
run;
```

```
proc sort data = tests;
  by name;
run;
proc sort data = hwk;
  by name;
run;
data all;
  merge tests hwk;
  by name;
run;
```

The SAS data set WORK.ALL will have a line for each student containing the student's name, his/her grades on three tests, and his/her grades on five homework assignments. Note that in order to match merge BY NAME, both data sets must be sorted in order BY NAME.

Since most learning is motivated by the necessity to solve a particular problem, we will now leave the general discussion of how to write SAS programs and proceed to consider how to use SAS to perform particular statistical analyses. Subsequent chapters show how to use SAS to produce the results obtained in Rao's corresponding chapters.

2

Theoretical Distributions

2.1 Introduction

Rao's Chapter 2, "Describing Statistical Populations," introduces several commonly used probability distributions which can be used to model statistical populations. In his Chapter 3, "Statistical Inference: Basic Concepts," Rao describes several more theoretical distributions which can be used to describe the (theoretical) behavior of statistics calculated from sample data. Statistical decisions are made based on probabilities, and most statistics textbooks contain tables which allow the user to look up probabilities associated with various values of a random variable with the specified probability distribution. Such tables are typically rather limited, giving only a few selected probabilities, because of space considerations.

In this chapter, we learn how to use SAS to look up probabilities or percentage points of the distributions discussed in Chapters 2 and 3 of Rao. Note that since these probability distributions are theoretical constructs, there will be no data entered in the SAS programs. In Chapter 3, "Observed Distributions," we will return to using SAS to organize and analyze data.

2.2 Probability Models of Statistical Populations

In Chapter 2, Rao introduces two discrete probability distributions, the *Binomial* and the *Poisson,* and the continuous *Normal* distribution. In this section, we demonstrate how to use SAS to compute probabilities for random variables having these distributions.

The Binomial Distribution

Example 2.16. In order to estimate the germination rate of a new seed variety, a plant breeder decided to plant six seeds in six identical pots and observe the number of germinating seeds. Let π denote the germination rate for the seed variety; that is, suppose that, on the average, we expect $100\pi\%$ of the planted seeds to germinate. Under the assumption that germination of each seed is independent of the other seeds, this is a binomial experiment with $n = 6$; trials and π is the probability that a seed will germinate. If Y is the number of germinating seeds, then Y is a binomial random variable with parameters $n = 6$ and π.

As an example, suppose that the germination rate is 70% ($\pi =0.70$). Then the probability that y out of $n = 6$ seeds will germinate is

$$f(y) = \frac{6!}{y!(6-y)!}(0.70)^y (1-0.70)^{6-y} \qquad y = 0, 1, 2, ..., 6.$$

For example, the probability of exactly two seeds germinating is ... 0.0595, whereas the probability of at least five seeds germinating is ... 0.4201.

Instead of computing the probabilities of exactly two and at least five of six seeds germinating either by hand or by looking in printed tables, we want to have SAS compute the probabilities for us. In Figure 2.1 is a SAS program to compute and print the desired probabilities.

The SAS command to compute binomial probabilities is PROBBNML. This command is executed in a data step. SAS will compute the probability, and this probability can be used in other calculations; however, SAS will not print the probability in an output file unless it is instructed to by the PRINT procedure.

Figure 2.1. Computing binomial probabilities

```
options ls=70 ps=54;
data binom;
  p1=probbnml(0.7,6,2) - probbnml(0.7,6,1);
  p2=1-probbnml(0.7,6,4);
run;
proc print data=binom;
  title 'Binomial Probabilities';
run;
```

The PROBBNML command computes *cumulative* binomial probabilities, that is,

$$P(Y \le y)$$

for y a value of the binomial variable Y. Thus, the command

```
probbnml(p,n,y)
```

gives the probability that there will be *y or fewer* successes in n trials, where the probability of success on any trial is π. In Figure 2.1, the probability of exactly 2 successes is found by computing the probability of 0, 1 or 2 and subtracting from it the probability of 0 or 1 successes. Similarly, the probability of 5 or 6 successes is obtained by subtracting the probability of 0, 1, 2, 3, or 4 successes from 1.

The output from this program is shown in Figure 2.2. Compare the results to those in Rao's Example 2.16.

Figure 2.2. Binomial probabilities

```
            Binomial Probabilities

      OBS        P1           P2

       1      0.059535     0.42018
```

The Poisson Distribution

Example 2.20. Refer to Example 2.16. Suppose that $n = 200$ seeds with a small germination rate, say 2%, were planted If the seeds germinate independently of each other, the random variable Y, the number of germinating seeds, will have a binomial distribution with $n = 200$ and $\pi = 0.02$. The probability that $Y = y$ can be calculated using the binomial probability function with $n = 200$ and $\pi = 0.02$ . . .Computation of $f(y)$ using [the binomial probability function] can be quite laborious. Since $n \geq 100$ is large and $n\pi = 4 < 5$, a Poisson approximation with $\lambda = n\pi = 200(0.02) = 4$ can be used to obtain a computationally simple expression for $f(y)$. Based on the Poisson approximation to the binomial, the relative frequency $f(y)$ ... is approximately equal to

$$f(y) = e^{-4}\frac{4^y}{y!}$$

so that the probability that six out of 200 seeds will germinate can be approximated as

$$\Pr\{Y = 6\} = e^{-4}\frac{4^6}{6!} = 0.1042.$$

The SAS command to compute Poisson probabilities is POISSON. Writing

```
poisson(λ,y)
```

computes the probability of *y or fewer* occurrences when the mean number of occurrences per interval of time or space is λ. As with the binomial probability command, then, to compute the probability of exactly six seeds germinating, subtract the cumulative probability of five or fewer from the cumulative probability of six or fewer. The program in Figure 2.3 produces the probability in Figure 2.4, which is the same value as obtained by Rao in Example 2.20.

Figure 2.3. Computing a Poisson probability

```
options ls=70 ps=54;
data poisson;
  p = poisson(4,6) - poisson(4,5);
run;
proc print data=poisson;
  title 'Poisson Probability';
run;
```

Figure 2.4. A Poisson probability from SAS

```
                    Poisson Probability

                   OBS        P

                    1      0.10420
```

The Normal Distribution

Example 2.23. Suppose that the yield of a new variety of corn is normally distributed with mean $\mu = 100$ bushels per acre and standard deviation $\sigma = 20$ bushels per acre. What is the probability that the yield will be less than 60 bushels per acre? What is the probability of an observed yield between 90 and 130 bushels per acre?

The SAS function PROBNORM(z) gives the probability that a standard normal random variable has a value *less than or equal to z.* To find the probability that the corn yield in Example 2.23. will be less than (or equal to) 60 bushels, we compute that 60 bushels is two standard deviations below the mean yield:

$$z = \frac{60-100}{20} = -2.0.$$

PROBNORM(-2) gives the probability of a yield this low or lower. To find the probability of a yield between 90 and 130 bushels per acre, we subtract the probability of 90 or less from the probability of 130 or less, as shown in the program in Figure 2.5.

Figure 2.5. Program to compute normal probabilities

```
options ls=70 ps=54;
data norm;
   p1 = probnorm(-2);
   p2 = probnorm(1.5) - probnorm(-0.5);
run;
proc print data=norm;
   title 'Normal Probabilities';
run;
```

First, 90 bushels is 0.5 standard deviations below average:

$$z = \frac{90 - 100}{20} = -0.5,$$

and 130 bushels is 1.5 standard deviations above average:

$$z = \frac{130 - 100}{20} = 1.5.$$

The probabilities computed by SAS are given in Figure 2.6.

Figure 2.6. Normal probabilities from SAS

Normal Probabilities		
OBS	P1	P2
1	0.022750	0.62466

Example 2.24. The serum iron content (mg/100 ml) in a population of subjects is known to be normally distributed with a standard deviation of σ = 5 mg/100 ml. Also, 33% of the population is known to have a blood serum of at least 115 mg/100 ml. We are interested in determining the mean serum level for the population.

In this example, the probability of exceeding 115 is given as 0.33. What is required is to determine how many standard deviations 115 is above the mean of the distribution; that is, to what z-value does 115 correspond? Percentiles (z-values) of the standard normal distribution are given by the SAS function PROBIT. PROBIT(p) gives the z-value such that 100p% of the area under the standard normal distribution lies to the left of z.

In Example 2.24, if the probability of exceeding 115 is 33%, then 67% of the area under the curve lies to the left of 115. To find out what z-value 115 corresponds to, write

$$probit(0.67)$$

as in Figure 2.7.

Figure 2.7. SAS program using PROBIT to find a z-value

```
options l70 ps=54;
data z;
   z = probit(0.67);
run;
proc print data = z;
   title 'Z-Value for Y = 115';
run;
```

Figure 2.8 indicates that the value 115 is 0.44 standard deviations above the mean of the population. Since the standard deviation is known to be 5, we have that

$$115 = 0.44(5) + \mu$$

or

$$\mu = 112.8,$$

as in Rao's Example 2.24.

Figure 2.8. Output from the PROBIT function

```
                      Z-Value for Y = 115

                 OBS              Z

                  1           0.43991
```

  Example 2.27. Soil properties such as bulk density, organic matter content, clay content and soil water content are generally characterized by a normal distribution... However, flow related properties such as air permeability, saturated hydraulic conductivity, and soil-water flux, have been reported to be log-normally distributed.

  In case a logarithmic transformation needs to be made to data in order for the distribution to appear normal, the following SAS functions can be employed in a data step.

  The command

$$LOG(Y)$$

computes the natural (base e) logarithms of the values of the variable Y. The inverse function is

$$EXP(X)$$

which raises e to the power X.

  In case base 10, or common logarithms are needed, the SAS function is

$$LOG10(Y)$$

and the inverse function is, of course,

$$10**X.$$

2.3 Sampling Distributions of Statistics

In Chapter 3, Rao introduces the notion of sampling distributions that describe the theoretical behavior of statistics computed from random samples.   The most commonly used sampling distributions are the *t-distribution* that models the behavior of sample means and the difference between two sample means, the *Chi-Square distribution* that describes the behavior of sample standard deviations, and the *F-distribution* that is used to model the behavior of the ratio of two sample variances.

Statistics textbooks, including the Rao text, commonly provide tables of these distributions, which can be used to look up probabilities or percentage points of the distributions. Since all three of these distributions are continuous, the tables are limited, giving only selected values. Thus, it is usually possible to make a statement like, "The probability is greater than 0.025 but less than 0.05," using the tables, but a more precise statement is not possible. Computers, however, are not subject to the same limitations as tables, and SAS can compute probabilities exactly and "look up" the percentiles or critical values corresponding to such probabilities. Usually, however, no more than four decimal places are displayed for probabilities in SAS output.

The t-Distribution

Example 3.10. Let's find the 0.025-level and 0.975-level critical values of a *t*-distribution with $v = 12$ degrees of freedom.

The PROBT (not to be confused with PROBIT) function in SAS computes probabilities from the *t*-distribution, and the TINV function finds percentiles of the *t*-distribution. In Example 3.10, the problem is to find the percentiles. Writing

```
tinv(p,v)
```

finds the *t*-value in a distribution with ν degrees of freedom such that 100p% of the area under the curve lies to the left. The program in Figure 2.9 finds *t*-values that cut off 2.5% of the area under the lower and upper tails of a *t*-distribution with 12 degrees of freedom. By symmetry, the two values are the same distance from the mean, but in opposite directions, as seen in Figure 2.10 and Rao's Example 3.10.

Figure 2.9. Finding percentiles of the *t*-distribution

```
options ls=70 ps=54;
data t;
   t1 = tinv(0.025,12);
   t2 = tinv(0.975,12);
run;
proc print data = t;
   title 'Percentage Points of a t-
       Distribution';
run;
```

Figure 2.10. Percentiles for a *t*-distribution

	Percentage Points of a t-Distribution	
OBS	T1	T2
1	-2.17881	2.17881

As an example of using SAS to compute probabilities for *t*-values, consider

$$\Pr(-1.86 \leq t(8) \leq 1.86).$$

Figures 2.11 and 2.12 show how the PROBT function can be used to compute this probability. As with the other distributions we have considered in this chapter, the *t*-distribution in SAS is cumulative, so we compute the probability of a value less than or equal to 1.86 and subtract the probability of a value less than −1.86.

Figure 2.11. Computing probabilities using the *t*-distribution

```
options ls=70 ps=54;
data tprob;
  p = probt(1.86,8) - probt(-1.86,8);
run;
proc print data = tprob;
  title 'Example 3.5.3';
run;
```

Figure 2.12. Probabilities for *t*-values

```
                      Example 3.5.3

                    OBS        P

                     1      0.90007
```

The Chi-Square Distribution

Example 3.11. Let us find the 0.05 and 0.95 critical values of a χ^2-distribution with $v = 12$ degrees of freedom.

The PROBCHI function in SAS computes Chi-Square probabilities; the CINV function finds percentiles. The program in Figure 2.13 shows how to find the 0.05 and 0.95 percentiles for a Chi-Square with 12 degrees of freedom using CINV(p,v). As with the other distributions we have considered in this chapter, the Chi-Square distributions are *cumulative*. For purposes of illustration, Figure 2.13 also shows how to use PROBCHI(y,v) to find p, the probability that a Chi-Square value with 10 degrees of freedom exceeds the value 18.23. Output from the program in Figure 2.13 is shown in Figure 2.14.

Figure 2.13. A program to find Chi-Square probabilities and percentiles

```
options ls=70 ps=54;
data chisq;
  x1 = cinv(0.05,12);
  x2 = cinv(0.95,12);
  p = 1 - probchi(18.23,10);
run;
proc print data=chisq;
  title 'Chi-Square Probability and
     Percentage Points';
run;
```

Figure 2.14. Chi-Square probabilities and percentiles

```
        Chi-Square Probability and Percentage Points

     OBS        X1         X2          P

      1      5.22603    21.0261    0.051206
```

The F-Distribution

Example 312. Let us determine the 0.05- and 0.95-level critical values of an $F(5,12)$-distribution.

The SAS function PROBF(y,δ,v) computes the probability that a random variable whose distribution is F with numerator degrees of freedom δ and denominator degrees of freedom v will take on a value less than or equal to y. The inverse function, FINV(p,δ,v) finds the $100p$ percentile of the distribution.

The program in Figure 2.15 finds the percentage points requested in Rao's Example 3.12 and also finds p, the probability that an F with 1 and 19 degrees of freedom exceeds the value 20.4. Output from this program is in Figure 2.16.

Figure 2.15. Program using PROBF and FINV functions

```
options ls=70 ps=54;
data f;
  f1 = finv(0.05,5,12);
  f2 = finv(0.95,5,12);
  p = 1 - probf(20.4,1,19);
run;
proc print data = f;
  title 'F Probability and Percentage
      Points';
run;
```

Figure 2.16. Probability and percentiles from an *F*-distribution

```
        F Probability and Percentage Points

   OBS       F1        F2          P

    1      0.21378   3.10588    .00023610
```

3

Observed Distributions

3.1 Introduction

Chapters 2 and 3 of Rao's text discuss techniques for describing populations and samples of observed data. In this chapter, we look at several SAS procedures for summarizing data.

3.2 The CHART Procedure

Example 3.2. The disease-free survival (DFS) time of a treated cancer patient is defined as the length of elapsed time between the time at which the patient goes into remission (becomes free of cancer) and the time at which the patient relapses (cancer recurs). The following data show the DFS times, in months, of a random sample of $n = 20$ breast cancer patients.

A basic method for obtaining an overview of a set of data, whether the data are considered a population or a sample, is to construct a *frequency distribution* for the data. A frequency distribution may be either a table or a graph, showing the different values in the data set, along with a count of how often or what percentage of the time each value occurs. If there are only a few different values in the set, then each individual value can be listed. On the other hand, if there are very many different values in the set, then in order to summarize the data effectively, these values need to be grouped into meaningful classes, intervals, or categories. This is so whether the data are qualitative or quantitative.

The CHART procedure in SAS can be used to produce frequency tables and graphs for both qualitative and quantitative data. We illustrate its use with the disease-free survival time data, which have been entered into a data file named DFS.DAT. This data file is shown in Figure 3.1. Note that in Figure 3.1, as in Rao's Example 3.2, the data are entered in order of magnitude. However, this is not a requirement of the CHART procedure.

Figure 3.1. The data file DFS.DAT

```
 2   2   3   6   6   7   9  11  12  12  13  15  19  19  21
23  24  30  44  45
```

Figure 3.2 shows a SAS program to produce a frequency distribution for these data. Actually, the program in Figure 3.2 produces two graphs and one table. In the CHART procedure, one can request either a vertical or a horizontal bar chart or histogram. If the horizontal chart is requested, a tabular summary is also produced. To request a vertical bar chart, simply write

```
proc chart;
   vbar months;
```

Figure 3.2. SAS program for vertical and horizontal bar charts

```
options ls=70 ps=54;
data a;
   infile 'DFS.dat';
   input months @@;
run;
proc chart;
 hbar months;
run;
proc chart;
   vbar months;
run;
```

where VBAR is an abbreviation for "vertical bar"; the command for horizontal bars is HBAR. Figures 3.3 and 3.4 show the output produced by this program.

Figure 3.3. Horizontal bar chart for DFS data

```
MONTHS                                       Cum.            Cum.
Midpoint                              Freq   Freq  Percent  Percent
         |
      5  |****************************    7     7   35.00    35.00
         |
     15  |***************************     7    14   35.00    70.00
         |
     25  |************                    3    17   15.00    85.00
         |
     35  |****                            1    18    5.00    90.00
         |
     45  |********                        2    20   10.00   100.00
         |
         ----+---+---+---+---+---+---+
             1   2   3   4   5   6   7

                   Frequency
```

Figure 3.3 shows the horizontal bar chart and tabular frequency distribution for the soil-water flux data. SAS automatically chooses to group the values into 5-7 intervals, depending on how large the data set is. The midpoints of the intervals are printed. The midpoints in Figure 3.3 correspond to intervals 0-10, 10-20, ..., 30-40. Frequencies in the intervals are represented visually by rows of asterisks, where, in this case, four asterisks represent one observation. Additionally, a table of frequencies, percents and cumulative frequencies and percents is produced.

The vertical bar chart is shown in Figure 3.4. Note that there is no accompanying tabular frequency distribution.

Since the CHART procedure automatically divided the data into approximately seven intervals, it might happen that the resulting distributions are less than satisfactory, since the automatically chosen intervals are hard to interpret. Two options in the CHART procedure that, allow you to control how the intervals are formed are the DISCRETE option and the MIDPOINTS= option. Writing

Figure 3.4. Vertical bar chart for DFS data

```
Frequency

7 +          *****       *****
  |          *****       *****
  |          *****       *****
  |          *****       *****
  |          *****       *****
6 +          *****       *****
  |          *****       *****
  |          *****       *****
  |          *****       *****
  |          *****       *****
5 +          *****       *****
  |          *****       *****
  |          *****       *****
  |          *****       *****
  |          *****       *****
4 +          *****       *****
  |          *****       *****
  |          *****       *****
  |          *****       *****
  |          *****       *****
3 +          *****       *****       *****
  |          *****       *****       *****
  |          *****       *****       *****
  |          *****       *****       *****
  |          *****       *****       *****
2 +          *****       *****       *****                   *****
  |          *****       *****       *****                   *****
  |          *****       *****       *****                   *****
  |          *****       *****       *****                   *****
  |          *****       *****       *****                   *****
1 +          *****       *****       *****       *****       *****
  |          *****       *****       *****       *****       *****
  |          *****       *****       *****       *****       *****
  |          *****       *****       *****       *****       *****
  |          *****       *****       *****       *****       *****
  ----------------------------------------------------------------
             5           15          25          35          45

                        MONTHS Midpoint
```

```
proc chart;
    vbar y/discrete;
```

will produce a bar chart with a separate bar for every different value of *y*. That would not be a very effective summarization for the DFS data, since it would produce 16 separate bars. The DISCRETE option is useful when there are not many different numerical values in the data set. (If the variable is a character variable, the DISCRETE option is automatically invoked.)

The MIDPOINTS option allows you to define intervals as you like them. For example, to produce Rao's Table 3.1, write

```
proc chart;
    hbar months/midpoints = 2.5 to 47.5 by 5;
```

This produces the horizontal bar chart and table in Figure 3.5.

Figure 3.5. Horizontal bar chart with specified midpoints

```
   MONTHS                                          Cum.              Cum.
  Midpoint                              Freq  Freq    Percent    Percent
            |
     2.5    |***************              3      3     15.00      15.00
            |
     7.5    |********************         4      7     20.00      35.00
            |
    12.5    |********************         4     11     20.00      55.00
            |
    17.5    |***************              3     14     15.00      70.00
            |
    22.5    |***************              3     17     15.00      85.00
            |
    27.5    |                             0     17      0.00      85.00
            |
    32.5    |*****                        1     18      5.00      90.00
            |
    37.5    |                             0     18      0.00      90.00
            |
    42.5    |*****                        1     19      5.00      95.00
            |
    47.5    |*****                        1     20      5.00     100.00
            |
            -----+----+----+----+
                 1    2    3    4
               Frequency
```

Example 2.1. Identifying the distribution of the pest population is a key step in any pest management program. Suppose that there are five types of pests in a given region. The population of pest types can be considered as a collection of measurements with five distinct values: 1, 2, 3, 4 and 5. Each measurement represents one type of pest.

Example 2.7......Let $f(y)$ denote the relative frequency of pest type y, where $y = 1, 2, 3, 4, 5$. A possible probability distribution for such a population is given in Table 2.2.

The program in Figure 3.2 uses raw data as input. However, sometimes you have access to data which have already been organized into a frequency table, and you would like to produce a bar chart corresponding to the table. The CHART procedure can accept a frequency table as input data. Figure 3.6 shows how to input the frequency distribution in Rao's Table 2.2.

Figure 3.6. Inputting a frequency table to the CHART procedure

```
options ls=70 ps=54;
data a;
 input pest pct;
lines;
1 16
2 18
3 10
4 46
5 10
;
proc chart data=a;
   hbar pest/freq=pct type=percent discrete;
run;
```

First, note that a case, or data line, is a line in the frequency distribution. For each category, the pest type and the percentage of the population of that type are entered. Note that percentages or counts may be entered, but this program won't work if relative frequencies, i.e., decimal fractions, are entered. In order to tell CHART that data are already organized into a frequency distribution, the FREQ option identifies the frequency or percentage variable. The DISCRETE option, as mentioned above, prevents CHART from dividing the interval from 1 to 5 into 7 categories, but instead specifies that each value of PEST is a separate category. This is necessary because although pest type is actually a categorical variable, it was entered as a numerical variable in Figure 3.6. TYPE=PERCENT will result in the frequency axis of the graph being labeled as percent, rather than as frequency. The output from this program is shown in Figure 3.7.

Figure 3.7. The CHART procedure for data entered as a frequency table

```
   PEST                                    Cum.                   Cum.
                               Freq   Freq   Percent   Percent
          |
     1    |********                 16     16    16.00     16.00
          |
     2    |*********                18     34    18.00     34.00
          |
     3    |*****                    10     44    10.00     44.00
          |
     4    |**********************    46     90    46.00     90.00
          |
     5    |*****                    10    100    10.00    100.00
          |
          -----+----+----+----+---
              10   20   30   40

                 Percentage
```

The CHART procedure can do much more than is discussed here. For example, if we had another 20 observations of disease-free survival time for a different drug, CHART could produce two side-by-side bar charts so the distributions for the two drugs could be compared visually. To explore the capabilities of the CHART procedure, refer to Chapter 9 of *SAS Procedures Guide, Version 6, Third Edition.*

3.3 The UNIVARIATE Procedure

In the previous section, we saw that the CHART procedure could be used to obtain either a tabular or a graphical frequency distribution for a set of data. In Chapter 1, we used the MEANS procedure to obtain various simple statistics for a set of data. There are several other SAS procedures that can be useful in summarizing data on a single variable, but the most complete information for a single variable is given by the UNIVARIATE procedure. Figure 3.8 shows a SAS program for using the UNIVARIATE procedure to summarize the DFS data.

Figure 3.8. SAS program to analyze DFS data using the UNIVARIATE procedure

```
options ls=70 ps=54;
data a;
   infile 'DFS.dat';
   input months @@;
run;
proc univariate plot;
   var months;
run;
```

First, note that although Figure 3.1 and Rao's Example 3.2 show the data arranged in numerical order, this is not a requirement of the UNIVARIATE procedure (or of any SAS procedure). To obtain summary statistics for these data, all that is necessary is to specify that the UNIVARIATE procedure be performed and to name the variable for which the analysis is required. The PLOT option in the UNIVARIATE procedure adds several visual displays to the output. Figure 3.9 shows the portion of the output produced without the PLOT option, and the additional output produced by the PLOT option is shown in Figure 3.10.

The UNIVARIATE output is presented in three sections: Moments, Quantiles, and Extremes. Under "Moments" we first see that the data set consists of $N = 20$ observations. In this data set, each observation is given a *weight* of 1, so that Sum Wgts = 20 as well. (Refer to Chapter 42: The UNIVARIATE Procedure, in *SAS Procedures Guide, Version 6, Third Edition* for instructions on how to incorporate weights.) The mean of the 20 values is 16.2555, and the sum is 325.11. The standard deviation and variance are 12.32328 and 151.8631, respectively.

Skewness and *kurtosis* are two measurements describing the shape of the distribution. Without going into much detail, skewness is departure from symmetry. For a perfectly symmetric distribution, the skewness measure is zero. If the values to the right of the mean are more spread out than are the values below the mean, the distribution is said to have *positive skewness* and the skewness measure will be a positive number. *Negative skewness* indicates that there is a "tail" on the left end of the distribution. The farther

Figure 3.9. UNIVARIATE analysis of DFS data

```
                      Univariate Procedure

Variable=MONTHS


                           Moments

          N               20   Sum Wgts          20
          Mean       16.2555   Sum           325.11
          Std Dev   12.32328   Variance     151.8631
          Skewness   1.17205   Kurtosis     0.990075
          USS       8170.225   CSS          2885.399
          CV        75.80988   Std Mean     2.755568
          T:Mean=0  5.899146   Pr>|T|         0.0001
          Num ^= 0        20   Num > 0           20
          M(Sign)         10   Pr>=|M|        0.0001
          Sgn Rank       105   Pr>=|S|        0.0001



                      Quantiles(Def=5)

          100% Max     45.02        99%      45.02
           75% Q3      22.105       95%     44.465
           50% Med     12.66        90%     37.025
           25% Q1       6.655       10%      2.92
            0% Min      2.23         5%      2.32
                                     1%      2.23

          Range        42.79
          Q3-Q1        15.45
          Mode          2.23



                         Extremes

        Lowest    Obs       Highest    Obs
         2.23(     1)        23.03(     16)
         2.41(     2)        23.61(     17)
         3.43(     3)        30.14(     18)
         6.3(      4)        43.91(     19)
         6.38(     5)        45.02(     20)
```

the value is from zero, the more highly skewed the distribution. The larger (positively) the value of the kurtosis measure, the "heavier" the tails of the distribution , i.e., the data contain some values that are quite a bit above or below the mean of the set. A normal distribution (mentioned in Chapter 2 and in Rao's Section 2.6, has kurtosis measure of zero, and serves as a reference distribution. Distributions with kurtosis less than zero have shorter tails than a normal distribution. For both skewness and kurtosis

Figure 3.10. Plots from the UNIVARIATE Procedure

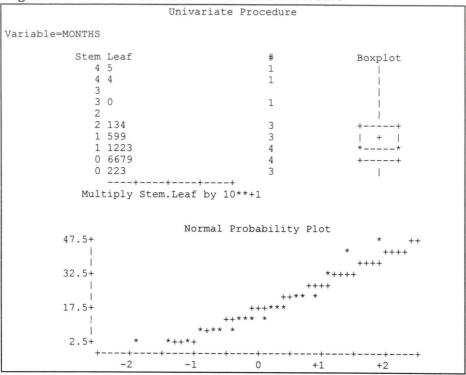

measures, experience in looking at histograms of your data gives you a feeling for what the numerical values mean.

The next line under Moments in the UNIVARIATE output gives "uncorrected" and "corrected" sums of squares, USS and CSS, respectively. The uncorrected sum of squares is simply the sum of the squares of the values,

$$USS = \Sigma y_i^2$$

while the corrected sum of squares is the sum of squared deviations of the values from their mean, or the sum of squares of the values, "corrected for the mean":

$$CSS = \Sigma(y_i - \bar{y})^2.$$

CV stands for "coefficient of variation", and is the standard deviation expressed as a percentage of the mean value:

$$CV = \frac{s}{\bar{y}} \times 100.$$

You may find this statistic useful in evaluating whether the standard deviation for a set of data is "small" or "large", since the CV expresses the size of the standard deviation relative to the average size of the values themselves.

In Section 3.5 of Rao, the standard deviation of the sampling distribution of the sample mean (the standard error of the mean) was seen to be

$$\sigma/\sqrt{n}$$

where σ is the population standard deviation. Std Mean = 2.755568 from the printout, is

$$s/\sqrt{n}$$

where s is the standard deviation of the sample.

The next line shows the value of the t-statistic for testing the null hypothesis that the mean of the population is zero, and the p-value for this test. Rao discusses hypothesis tests and p-values in Section 3.8; we will take up this topic in Chapter 4.

In some applications, it is useful to know the number of nonzero data points and the number of positive data points. These are given by

```
Num ^= 0
```

and

```
Num > 0,
```

respectively.

Finally, the last two lines under the Moments section of the UNIVARIATE printout give the value of the test statistic and the p-value for the Sign Test and the Signed Ranks Test. The Sign Test will be discussed in Chapter 6: Categorical Data; the Signed Ranks Test will be discussed in Chapter 5: Ordinal Data.

The next section of UNIVARIATE output is the "Quantiles" section. In the left column are given the maximum and minimum values and the *quartiles* of the distribution. The quartiles divide the data into four sections, in order of magnitude. The smallest value in the data set is 2.23; one-fourth of the values (in this case, 5 values) are less than or equal to 6.655, half of the values are less than or equal to 12.66, three-fourths of the values are no greater than 22.105, and the largest value is 45.02.

Below the quartiles are given the *range*, the distance between the largest and the smallest values,

```
Range = 45.02 - 2.23 = 42.79
```

and the *interquartile range* (IQR), which is the distance (difference) between the first and third quartiles,

```
IQR = Q3 - Q1 = 22.105 - 6.655 = 15.45.
```

The IQR is the distance spanned by the middle-sized half of the data. It is also known as a 25% *trimmed range*: the largest 25% of the data are "trimmed off", as are the smallest 25%, and the IQR is the range spanned by what is left.

The *mode* is the value which occurs the most frequently in the data. As is the case in this data set, when there is no mode, the smallest data value is given. (In the case of more than one mode, the smallest one is given.) For measurements made on a continuous variable, it is quite possible that each value will occur only once, so that there will be no mode.

The right column under Quantiles gives the 1%, 5%, and 10% smallest and largest values; that is, for example, 5% of the values are less than or equal to 2.32, and 5% of the values are greater than or equal to 44.465. There are several different ways to compute percentiles of a distribution. SAS provides five different algorithms in the UNIVARIATE procedure. The heading

```
(Def=5)
```

on this section indicates that these quantities are computed using method number 5. The interested reader may refer to Chapter 42 of *SAS Procedures Guide*.

Had this data set contained missing values, the UNIVARIATE procedure would have also produced a count of how many values were missing and what percentage of the data set consisted of missing values.

Under the section labeled "Extremes" are the five smallest values in the data set and the five largest values. These largest and smallest values are identified by their observation numbers. In this example, the data were arranged in order from smallest to largest in the input data set, so that the first one--Observation #1--was the smallest and the twentieth one was the largest. Use of an *ID statement* causes the values of the ID variable to be used to identify the extreme values. For example, suppose that the data set consists of names of students, in alphabetical order, and their grades on a test. Writing

```
proc univariate;
  var grade;
  id name;
```

will produce a univariate analysis of the grade distribution, using the students' names to identify the top five and the bottom five students. The ID statement can be used in any procedure in which SAS produces a full or partial list of cases.

Referring back to the program in Figure 3.8, notice the PLOT option in the

```
proc univariate plot;
```

command. Output resulting from the PLOT option is shown in Figure 3.10. Three plots are produced: a stem-and-leaf diagram, and box-and-whisker plot, and a normal probability plot.

A *stem-and-leaf* plot is like a bar graph, except that it shows more detail. SAS automatically groups the data into approximately 20 intervals, called "stems". In this example, natural intervals would be, say, 0-10, 10-20, . . . 40-50, resulting in five intervals. Since this is quite a bit fewer than 20, SAS divides each interval into two, resulting in intervals "low units", "high units", "low teens", "high teens", and so on. The "stems" are taken to be the *first digits*--in this case, the tens digits--of the intervals: 0, 10, 20, 30, 40. Note the legend at the bottom of the diagram:

```
Multiply Stem.Leaf by 10**+1.
```

The "leaves" on a stem-and-leaf diagram are the next digits after the stems. For example, the value 45.02 has stem 4 and leaf 5, since 45.02 is rounded to 45. Rounding the value 43.91 to the nearest whole number gives 44, and this value has stem 4 and leaf 4, and so on.

In this example, in order to get as close as it can to 20 stems, UNIVARIATE split each stem in two. In other cases, it might be required to *combine* adjacent stems in order to obtain a plot with approximately 20 stems. For example, the soil flux data in Figure 3.11 (described in Rao's Example 2.27) produces the stem-and-leaf diagram in Figure 3.12. Since that data set had only 20 observations, instead of producing a plot with approximately 20 stems, UNIVARIATE combined stems. On the 6 stem,

Figure 3.11. Soil-water flux data from Rao's Example 2.27.

```
0.3780 0.5090 0.6230 0.6860 0.7350 0.7500
0.7520 0.8690 0.8890 0.8890 0.8990 0.9370
0.9820 1.0220 1.0370 1.0880 1.1230 1.2060
1.3310 1.4230
```

Figure 3.12. Stem-and-leaf diagram for soil flux data

```
    Stem Leaf                       #          Boxplot
     14 2                           1             |
     12 13                          2             |
     10 2492                        4          +-----+
      8 799048                      6          *--+--*
      6 29455                       5          +-----+
      4 1                           1             |
      2 8                           1             |
        ----+----+----+----+
     Multiply Stem.Leaf by 10**-1
```

for example, one would read the data points as 0.62, 0.69, 0.74, 0.75, and 0.75.

Next to the stem-and-leaf diagram, UNIVARIATE prints a *box-plot* (sometimes called a "box-and-whiskers plot") of the data, using the same scale as is given for the stem-and-leaf diagram. Referring back to Figures 3.9 and 3.10, we see that the ends of the box are at the first and third quartiles of the data (about 7 and 22, respectively) and the dashed line through the box is at the median, (approximately 12). The plus sign inside the box indicates the mean (about 15). The "whiskers" extend to the minimum value (approximately 2) and the maximum (about 45).

Points that are 1.5 IQRs beyond the ends of the box in either direction are called *inner fences;* points that lie a 3-IQR distance from the box are called *outer fences.* Data points that lie between the inner and outer fences are designated *mild outliers;* those which lie beyond the outer fences are *extreme outliers.* Should the data set contain any outliers, then instead of extending the whiskers to the minimum and maximum values, UNIVARIATE draws the whiskers to the inner fences, and then denotes mild outliers with the symbol "0" and extreme outliers by the symbol "*".

Side-by-side box-plots are an especially convenient way to compare distributions. For example, suppose there are four large sections of a freshman chemistry course, and you want to look at the grade distribution for each section and also to compare the sections. Writing

```
proc univariate plot;
   var grades;
   by section;
```

(after having sorted the data by section) will produce univariate analyses for each section separately, and then will produce side-by-side box-and-whisker plots. In such a plot, it is easy to compare the four sections in terms of average, variability, shape of distribution, and presence of unusually high or low grades.

Finally, the n*ormal probability plot* in Figure 3.10 provides a way to evaluate visually the degree to which the distribution of the data departs from a normal distribution. Refer to Section 8.6 of Rao for a complete discussion of how these plots are constructed and what they are used for. Briefly, the normal probability plot produced by the UNIVARIATE procedure graphs the percentiles of the data distribution on the vertical axis, and the percentiles of the standard normal distribution on the horizontal axis. If the data are approximately normally distributed, then a graph of the data percentiles against the normal percentiles will produce a straight line. This straight line is indicated by plus signs. The actual data percentiles plotted against normal percentiles are indicated by asterisks. If the data are approximately normally distributed, then, the asterisks will tend to coincide with the plus signs. When this is so, UNIVARIATE prints the asterisk. Thus, if a large number of plus signs are visible, this is an indication of departure of the data from a normal distribution, as is the presence of asterisks out of line with the plus signs.

The normal probability plot in Figure 3.10 indicates departure from normality primarily by the visibility of a lot of plus signs, and also by the several asterisks that are out of line with the plus signs. It is also possible to produce a formal test for departure from normality by including the

NORMAL option before, after, or instead of the PLOT option in the PROC UNIVARIATE command.

Comparing all three graphs helps you learn how to interpret them all. All three graphs in Figure 3.10 are indicating that these data show a mound-shaped, but right-skewed distribution, with no outliers.

Another option which is available in the UNIVARIATE procedure is the FREQ option. It is specified in the PROC UNIVARIATE command, either before, after, or instead of the PLOT command. This may be a useful option if the data set is not large and/or contains only a few distinct values. Output from this option produces a list of the data values in numerical order, with the frequency, percentage, and cumulative percentage corresponding to each distinct data value. If the data set contains many distinct values, the output from the FREQ option is lengthy and does not provide an effective summarization of the data. (See the output from the FREQ procedure in Section 3.4 below.)

As was the case in the CHART procedure, the UNIVARIATE procedure can accept input data that are in the form of a frequency distribution. For example, refer to the distribution of pest types in Figure 3.6 (Rao's Table 2.2). Writing

```
proc univariate plot;
    var pest;
    freq pct;
```

would produce a univariate analysis. Note that the command indicating frequency in the UNIVARIATE procedure is different from the one in the CHART procedure (see Figure 3.6.) Hopefully, future releases of SAS will make this consistent from one procedure to another.

3.4 The FREQ Procedure

The FREQ procedure can be used to produce one- or two-way frequency tables, or even larger ones. It is most effectively used when there are just a few discrete values of the variables to be tabulated, as each individual value is listed and its frequency, relative frequency, and cumulative frequency are given. This is the same sort of distribution that is produced using the FREQ option in the UNIVARIATE procedure.

Figure 3.13 shows a program using the FREQ procedure to summarize the DFS data of Figure 3.1. Note that the variable whose frequency distribution is to be computed is not denoted by VAR, but rather by TABLES in the FREQ procedure. The output in Figure 3.14 shows that because there are 16 different values in that data set, the FREQ procedure doesn't do much of a summarization. However, there might be instances in which you would want a table as in Figure 3.14 and would want to use either the FREQ procedure or the FREQ option in the UNIVARIATE procedure. The FREQ procedure is the most useful for contingency table (two-way frequency distribution, or cross-classification) analysis, as we will see in Chapter 6: Categorical Data.

Figure 3.13. SAS program using the FREQ procedure to summarize the DFS data

```
options ls=70 ps=54;
data a;
   infile 'DFS.dat';
   input months @@;
run;
proc freq;
   tables months;
run;
```

Figure 3.14. The FREQ procedure applied to DFS data

MONTHS	Frequency	Percent	Cumulative Frequency	Cumulative Percent
2	2	10.0	2	10.0
3	1	5.0	3	15.0
6	2	10.0	5	25.0
7	1	5.0	6	30.0
9	1	5.0	7	35.0
11	1	5.0	8	40.0
12	2	10.0	10	50.0
13	1	5.0	11	55.0
15	1	5.0	12	60.0
19	2	10.0	14	70.0
21	1	5.0	15	75.0
23	1	5.0	16	80.0
24	1	5.0	17	85.0
30	1	5.0	18	90.0
44	1	5.0	19	95.0
45	1	5.0	20	100.0

4

Inferences about Means and Variances of One and Two Populations

4.1 Introduction

This chapter introduces several SAS procedures for estimating and testing hypotheses about the means and variances of single populations. In addition, methods for using the SAS system to test hypotheses or make estimates concerning either the difference in the means of two populations or the ratio of the variances of two populations are illustrated.

All of the procedures provided by SAS for performing inferences on means assume that the standard deviations of the populations are unknown, and the sample standard deviations are used to estimate the population standard deviations.

4.2 Estimating a Population Mean

Example 3.27. In Example 3.14 we presented data on the percentage of nitrogen loss (N-loss) over a 16-week period when Urea + N-Serve (UN) was used as the fertilizer for sugarcane. In the following table we add data for N-loss when Urea (U) alone was used as the fertilizer.

These data are independent random samples of $n_1 = 15$ plots under fertilizer UN and $n_2 = 13$ plots under fertilizer U.

Example 4.6. Consider the N-loss data for fertilizer UN in Example 3.27. Assume that the percentage N-losses have a normal distribution, but the population standard deviation, σ, is not known. Let's compute a 95% upper confidence bound for μ, the mean percentage N-loss for fertilizer UN.

The MEANS procedure in SAS is used to compute various univariate descriptive measures. In addition, it can be used to compute one- and two-sided confidence intervals on the mean of the population from which the data are assumed to be a random sample, and to test hypotheses concerning the population mean. Figure 4.1 shows the data referred to in Example 3.27, and Figure 4.2 shows how to use the MEANS procedure to obtain default statistics and to obtain one- and two-sided confidence limits for the population mean. Note that the N-loss data are stored in a file named "nloss.dat."

Figure 4.1. N-loss data

```
UN 10.8 10.5 14.0 13.5 8.0 9.5 11.8 10.0 8.7
9.0 9.8 13.8 14.7 10.3 12.8 .
U 8.0 7.3 14.1 9.8 7.1 6.3 10.0 7.1 7.9 6.1
6.9 11.0 10.0 .
```

While a large number of univariate descriptive measures can be obtained using the MEANS procedure, the default output is governed largely by the LINESIZE option. If an option is used so that the output fits on 8 1/2 × 11-inch paper, then for each numerical variable in the data set or for only those numerical variables listed in the VAR (for VARiables) statement, the number of cases, mean, standard deviation, and smallest and largest values are given automatically, as is shown in the first part of Figure 4.3.

The command CLM in the MEANS procedure stands for "confidence limits for the mean". It produces two-sided 95% confidence limits for the population mean. The second page of the output shown in Figure 4.3 gives the upper and lower 95% confidence limits for μ as

9.9617 and 12.3316,

Figure 4.2. The MEANS procedure applied to N-loss data

```
options ls=70 ps=54;
data a;
  infile 'nloss.dat' missover;
  input fert $ @;
    do until (loss = .);
      input loss @;
      if loss ne . then output;
    end;
run;
proc means data=a;
  var loss;
  where fert = 'UN';
  title 'Default Output from the Means
      Procedure';
run;
proc means data=a clm;
  var loss;
  where fert = 'UN';
  title 'Two-Sided 95% Confidence Limits';
run;
proc means data=a uclm;
  var loss;
  where fert = 'UN';
  title 'Upper 95% Confidence Limit';
run;
```

respectively.

If one-sided limits are required, the commands are LCLM and UCLM, for lower and upper 95% confidence limits for the population mean, respectively. The results are given on the third page of the output shown in Figure 4.3. The results are identical to Rao's result in Example 4.6 we can conclude with 95% confidence that the mean N-loss is no more than 12.12%.

To change the confidence coefficient from the default 95%, specify the value for α in the PROC MEANS command. For example, to obtain a two-sided 99% confidence interval, write

```
proc means data=a clm alpha=0.01;
```

Figure 4.3. Output from MEANS procedure

```
                 Default Output from the Means Procedure              1

                    Analysis Variable : LOSS


      N          Mean        Std Dev        Minimum         Maximum
     -----------------------------------------------------------------
     15     11.1466667      2.1397152      8.0000000      14.7000000
     -----------------------------------------------------------------



                 Two-Sided 95% Confidence Limits                      2

                 Analysis Variable : LOSS

                 Lower 95.0% CLM   Upper 95.0% CLM
                 ---------------------------------
                       9.9617319         12.3316014
                 ---------------------------------


                 Upper 95% Confidence Limit                           3

                 Analysis Variable : LOSS

                           Upper 95.0% CLM
                           ----------------
                               12.1197413
                           ----------------
```

4.3 Hypothesis Tests on a Single Mean

Example 4.7. Refer to the N-loss data for fertilizer UN in Example 3.27. The results in Box 4.2 can be used to test a hypothesis about μ, the population mean N-loss. Suppose we wish to test H_0: $\mu \geq 13$ against. H_1: $\mu < 13$. If a test at the $\alpha = 0.05$ level is desired, the upper confidence interval $(-\infty, 12.12)$ computed in Example 4.6 can be used as described in Box 4.2. Since $\mu_0 = 13\%$ is outside the 95% UCI, we reject the null hypothesis at the $\alpha = 0.05$ level and conclude that the true mean percentage N-loss is less than 13%.

Alternatively, the test statistic approach can be used. With $\mu_0 = 13$, the calculated value of the test statistic is $t_c = \ldots -3.35$. The research hypothesis is accepted at the level $\alpha = 0.05$, because the calculated value of -3.35 for the test statistic, t, is less than $- t(14, 0.05) = - 1.761$. Thus, we conclude that the true mean N-loss is less than 13%. The probability of an error when arriving at such a conclusion is less than 5%.

Alternatively, the p-value of the test can be computed. For 14 degrees of freedom in Table C.2 (Appendix C) we see that the p-value is less than 0.005. Therefore, the null hypothesis can be rejected at any level ≥ 0.005.

In the last section, we showed how a confidence interval (one- or two-sided) on a population mean can be computed. Tests of hypotheses concerning the population mean can be answered with reference to these confidence intervals, as illustrated in Rao's Example 4.7, above.

To use the test statistic and/or the p-value approach to testing hypotheses in SAS, either the UNIVARIATE or the MEANS procedure can be used. A t-value and associated p-value for testing

$$H_0 : \mu = 0$$

versus

$$H_1 : \mu \neq 0$$

is part of the standard output from the UNIVARIATE procedure. Refer back to Figure 3.9, for example. In the "Moments" section of the UNIVARIATE output, the seventh line is

```
   T:Mean=0     28.60479              Pr>|T|        0.0001
```

This is read as the t-value for testing the null hypothesis that the population mean is zero is 28.6049, and the two-sided p-value associated with this t-value is essentially zero. That is, the sample mean $\bar{y} = 75.545$ is an estimated 28.6 standard deviations above zero, and the probability that a deviation this large from zero in either direction could have occurred as a result of sampling error alone is very small (less than 0.0001).

The same test of $H_0 : \mu = 0$ can be obtained using the MEANS procedure by requesting the *t*-value and its associated (two-sided) *p*-value, as in

```
proc means t prt;
   var y;
run;
```

However, (1) it is not often that one wants to test the null hypothesis that the population mean is *zero*, and (2) probably more often than not, a one-sided test is appropriate to the research hypothesis, instead of a two-sided one. Fortunately, it is easy to obtain the desired tests in both UNIVARIATE and MEANS by (1) performing the analysis on the *deviations from* μ_0, since if the values average μ_0, then their average deviation from μ_0 is zero, and (2) dividing the two-sided *p*-value by 2 to obtain a one-sided *p*-value.

Figure 4.4 shows a program to test

$$H_0 : \mu = 13$$

versus

$$H_1 : \mu \neq 13$$

using the N-loss data, with both the UNIVARIATE and the MEANS procedures. Note that in the data step, deviations of the measurements from 13.0 are computed, and that these deviations are the analysis variable for both procedures.

The analysis from the MEANS procedure is shown on page 1 of the output file in Figure 4.5 and the analysis from UNIVARIATE is on the second page of the output file. (Only the Moments section from UNIVARIATE is shown in Figure 4.5.) In both cases,

$$t = -3.35462,$$

Figure 4.4. Testing H$_0$: $\mu = 13$ using UNIVARIATE and MEANS

```
options ls=70 ps=54 pageno=1;
data a;
  infile 'nloss.dat' missover;
    input fert $ @;
      do until (loss = .);
        input loss @;
        if loss ne . then output;
      end;
run;
data b; set a;
   dev=loss-13;
   where fert = 'UN';
run;
proc means data=b t prt;
   var dev;
   title 'Testing a Mean Using the MEANS
       Procedure';
run;
proc univariate data=b;
   var dev;
   title 'Testing a Mean Using UNIVARIATE';
run;
```

which differs from Rao's result in Example 4.7 only because of rounding error. To use the "test statistic" approach, this *t*-value is compared to

$$-t(14,0.005) = -1.761$$

and the null hypothesis is rejected in favor of the one-sided alternative.

To use the *p*-value approach we find in both sets of results that

$$p = 0.0047$$

for the two-sided test, as indicated by the absolute value signs around T in Prob>|T|. Thus, to test

$$H_1 : \mu < 13,$$

Figure 4.5. Hypothesis tests from MEANS and UNIVARIATE

```
                Testing a Mean Using the MEANS Procedure            1

                        Analysis Variable : DEV

                          T          Prob>|T|
                    ----------------------
                      -3.3546190      0.0047
                    ----------------------


               Testing a Mean Using UNIVARIATE                      2

                         Univariate Procedure

Variable=DEV


                              Moments

            N               15   Sum Wgts            15
            Mean       -1.85333   Sum              -27.8
            Std Dev    2.139715   Variance      4.578381
            Skewness   0.332969   Kurtosis      -1.24354
            USS          115.62   CSS           64.09733
            CV         -115.452   Std Mean      0.552472
            T:Mean=0   -3.35462   Pr>|T|          0.0047
            Num ^= 0         15   Num > 0              4
            M(Sign)        -3.5   Pr>=|M|         0.1185
            Sgn Rank        -45   Pr>=|S|         0.0084
```

the p-value is $p = 0.0047/2 = 0.00235$. This is consistent with Rao's result, using limited tables, that $p < 0.005$.

4.4   Comparing Means of Two Populations: Independent Samples

Example 4.8. To determine whether the waste discharged by a chemical plant is polluting the local river, the river water was sampled at two locations--one upstream and one downstream from the discharge site. Independent water samples of sizes $n_1 = 10$ and $n_2 = 15$, respectively, were selected from the upstream and downstream locations. The concentration level (ppm) of a suspected chemical pollutant was determined in each water sample.

The TTEST procedure in SAS is used to test the null hypothesis that there is no difference in the means of two normally distributed populations, based on a comparison of the means of samples drawn *independently* from the two populations. The TTEST procedure assumes that the population standard deviations are unknown and must be estimated by the sample standard deviations. It provides *t*-statistics and *p*-values for both the case in which the population variances can be assumed to be equal, and for the case in which they cannot. In addition, the TTEST procedure provides a test on whether the sample variances differ significantly.

The program in Figure 4.6 performs the TTEST procedure; the UNIVARIATE procedure is also used in order to obtain side-by-side box-plots to supplement the TTEST output.

The TTEST procedure requires a CLASS statement, which names the variable whose values define the two groups whose means are to be compared. Note that the CLASS variable can have *only* two values. If it has more than two, the TTEST procedure will not run and you will be presented with a message, "The CLASS variable has more than two levels," in your log file. The procedure also requires that the VAR statement names the numerical variable whose means are to be compared for the two groups.

Figure 4.7 shows the side-by-side box-plots for the pollution data. These plots indicate that the average ppm appears to be greater for the downstream locations than for the upstream locations, but that there is quite a lot of variability in both locations. The amount of variability does not appear to be much different in the two locations, however. Both sample distributions are somewhat skewed to the right, casting a bit of doubt on whether the assumption of normal populations is tenable.

Figure 4.8 contains the output from the TTEST procedure. The arrangement of the output on the page and the headings are somewhat confusing to the first-time user. There are three parts of the output: (1) a table of statistics for each of the two groups, (2) two versions of the *t*-test of the equality of population means, and (3) a test for equality of population variances. These three sections are not very clearly separated.

Figure 4.6. SAS program to run the TTEST procedure

```
options ls=70 ps=54;
data a;
   infile cards missover;
   input loc $ @;
      do until (ppm = .);
         input ppm @;
         if ppm ne . then output;
      end;
lines;
up 24.5 29.7 20.4 28.5 25.3 21.8 20.2 21.0
21.9 22.2 -1
down 32.8 30.4 32.3 26.4 27.8 26.9 29.0 31.5
31.2 26.7
25.6 25.1 32.8 34.3 35.4 -2
run;
proc sort;
   by loc;
run;
proc univariate plot;
   var ppm;
   by loc;
run;
proc ttest;
   class loc;
   var ppm;
 run;
```

Looking first at the summary statistics for the two groups, we see that the analysis variable is PPM, and that there are 15 cases in the downstream site and 10 in the upstream location. The mean percent ppm in the two locations are 29.88 and 23.55, respectively, and the standard deviations are 3.33 and 3.36. These standard deviations measure variability in the individual readings of ppm; the standard errors of 0.86 and 1.06 estimate variability in the sample means. (Recall that these standard errors are calculated as $s/\sqrt{n}$.) The difference in the sample means is evaluated relative to the sizes of these standard errors. Do the sample means differ more than can be explained by sampling variability alone?

Figure 4.7. Side-by-side box-plots for river pollution study

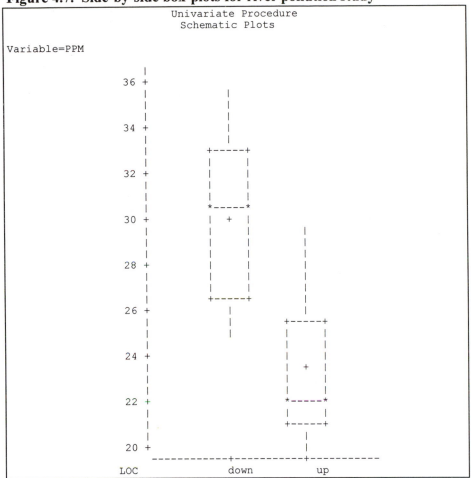

As explained in Rao's Box 4.3, if it can be assumed that the two populations have the *same* variances, then these standard errors (or their squares) can be pooled together to obtain an estimate of the common variance based on information from both samples. The difference in the sample means is then evaluated relative to this pooled standard deviation, using a *t*-statistic with $(n_1 - 1) + (n_2 - 1) = n_1 + n_2 - 2$ degrees of

Figure 4.8. TTEST analysis for river pollution study

```
                        TTEST PROCEDURE

Variable: PPM

LOC        N                  Mean            Std Dev         Std Error
--------------------------------------------------------------------------
down      15          29.88000000         3.32655978        0.85891404
up        10          23.55000000         3.35898463        1.06220421


Variances          T        DF     Prob>|T|
--------------------------------------------
Unequal        4.6339     19.3      0.0002
Equal          4.6433     23.0      0.0001

For H0: Variances are equal,  F' = 1.02    DF = (9,14)
                              Prob>F' = 0.9396
```

freedom. Otherwise, the *t*-statistic is formed with estimated standard error and degrees of freedom as in Rao's Box 4.4.

SAS gives both versions of the two-sample *t*-statistic in the output from the TTEST procedure. These are found in the middle section of Figure 4.8, under a heading beginning with "Variances." Keep in mind, however, that the *t*-test does not *test* variances--it tests means--but how it tests means depends on whether or not the population variances are equal. Thus, the two lines under "Variances" refer to whether the population variances are unequal or are assumed to be equal, respectively. The value of the *t*-statistic, assuming unequal population variances, is 4.6339, with 19.3 degrees of freedom. The value of the *t*-statistic, computed under the assumption of *equal* population variances, is $t = 4.6433$, with 23 degrees of freedom.

Depending upon whether or not the researcher thinks it is reasonable to assume that the variability in ppm is the same under both experimental conditions, (s)he will concentrate on one of the two lines in the middle portion of the TTEST output. If the truth be told, seldom, in practice, do the two versions of the *t*-statistic differ very much, and usually the same conclusion about whether or not to reject

$$H_0 : \mu_1 - \mu_2 = 0$$

in favor of

$$H_1: \mu_1 - \mu_2 \neq 0$$

is reached using both versions. In this case, *p*-values of $0.0002/2 = 0.0001$ and $0.0001/2 = 0.00005$ both lead us to conclude in favor of

$$H_1: \mu_1 - \mu_2 > 0,$$

where μ_1 is the mean ppm at the downstream location.

With a little effort, the TTEST procedure can be used to test

$$H_0: \mu_1 - \mu_2 = \Delta_0$$

where Δ_0 is some value other than zero. Suppose that there were some *a priori* reason to hypothesize that the mean ppm of the pollutant at the downstream location was expected to be 3 ppm greater than at the upstream site. That is,

$$H_0: \mu_1 - \mu_2 = 3$$

is to be tested. Note that this hypothesis can be written as

$$H_0: (\mu_1 - 3) - \mu_2 = 0$$

which gives us a clue as to how to get SAS to perform this test. In a data step create new values for the downstream readings by subtracting 3 from each value in the downstream group. This will lower the mean of the downstream group by 3 but will not affect the standard deviation. Then perform the TTEST procedure using these new values for the downstream group and the actual readings from the upstream samples. If the data are read in as in the program in Figure 4.6, then the following code will create the new variable:

```
data b; set a;
   if loc = 'up' then ppm = ppm -3;
   else ppm = ppm;
run;
proc ttest;
   class loc;
   var ppm;
run;
```

Note: When the variable name is the same on both sides of the equation, as in

$$ppm \; = \; ppm \; - \; 3;$$

the whole equation can be replaced by just the expression on the right-hand side. Thus, the IF...THEN statement above could be written

```
if loc = 'up' then ppm - 3;
```

Perhaps in a forthcoming edition, SAS will provide for interval estimates of $\mu_1 - \mu_2$ in conjunction with the TTEST procedure, but at present, if such an estimate is desired, you will have to compute it yourself, according to the formulas in Rao's Boxes 4.3 and 4.4. At least, the TTEST procedure gives you values for n_1, n_2, $\bar{y}_1$, $\bar{y}_2$, $s_1/\sqrt{n_1}$, and $s_2/\sqrt{n_2}$ to use in the formulas. Alternatively, the MEANS command in the GLM or ANOVA procedures, to be discussed in Chapter 8, can be used to find confidence limits for $\mu_1 - \mu_2$.

The last two lines of the TTEST output show the test of whether the two populations have unequal variances. We will return to this output in Section 4.6.

4.5 Comparing Means of Two Populations: Paired Samples

Example 4.10. Measurements of cell fluidity were made in 20 dishes of pulmonary artery cells from ten dogs. Cell samples from each dog were randomly divided into two dishes; a randomly selected dish from each pair was randomly assigned to an oxygen (O_2) treatment and the other to a control treatment consisting of no O_2 exposure.

Suppose we are interested in comparing μ_2, the mean fluidity of cells exposed to O_2 with μ_1, the mean for untreated cells. Methods for comparing the means of two populations that use data from independent samples are not appropriate because the measurements from dishes within a pair are unlikely to be independent. However, the sample of ten differences can be regarded as a random sample of size n = 10 from a population with mean $\mu_D = \mu_1 - \mu_2$.

Example 4.11. On the basis of the data in Example 4.10, let's construct a 95% confidence interval for μ_D, the difference in the true mean cell fluidity of treated and untreated cells.

Sometimes, there is a danger that preexisting characteristics of the subjects (experimental units, sampling units, cases) are likely to have an influence on the subjects' responses to an experimental treatment or observational variable. If two groups are to be compared, then it is desirable to ensure that the two groups start out being equal with respect to this preexisting characteristic. Otherwise, there is no way to tell whether any observed difference in response was due to differences in the treatments or only to differences in the preexisting characteristic. The process of making sure that the two groups are initially comparable on this characteristic is referred to as *controlling for* this characteristic.

One way to control for a preexisting characteristic is to *match* or *pair* subjects who have similar values on the characteristic. Then each member of a pair is randomly assigned to one of two experimental or observational groups. A variation on this theme is to apply each experimental treatment to, or to make an observation under each condition on, all subjects in the study; this is called *repeated measures*. In either case, the difference in observations under two conditions by the two members of a pair *cannot* reflect differences in the preexisting condition, but must reflect differences in the observational or experimental groups.

Forming the differences in responses for each member of a pair reduces a two-sample problem down to a one-sample problem: we concentrate not on the actual measurements made, but on the sample differences. To test hypotheses about or make estimates of the mean difference, the one-sample methods of Section 4.2 can be used.

Figure 4.9 shows a program that computes the differences in each pair of responses and then uses the MEANS procedure to obtain the sample size, sample mean, and sample variance, and then form a two-sided 95% confidence interval on μ_D, the population mean of the differences, and to compute the *t*-statistic and the associated *p*-value for testing

$$H_0 : \mu_D = 0.$$

Note carefully that the data for a paired sample *t*-test is entered quite differently from the way it is done for a *t*-test based on independent samples. In Figure 4.9, a case is a *pair* of observations, and that we have two response variables measured on each case. Entering data like this enables us to form the differences for each pair of response measurements. Contrast this with the way the data are entered in Figure 4.6. There is no logical way to match up a reading under the upstream location with one at the downstream site. As a matter of fact, there are not even the same number of readings at the two sites. For independent samples as in Figure 4.6, there are $n_1 + n_2$ cases, and one of the two variables measured on each case is the response and the other is the group designation.

The output from the program in Figure 4.9 is shown in Figure 4.10. The results are the same as in Rao's Example 4.11, except for the signs on the differences, the sample mean difference, the confidence limits, and the *t*-value. This difference in signs is a result of the differences in Figure 4.9

Figure 4.9. A program to estimate and test an hypothesis about μ_D

```
options ls=70 ps=54;
data a;
  input ctrl oxy;
  diff = oxy - ctrl;
lines;
0.308 0.308
0.304 0.309
0.305 0.305
0.304 0.311
0.301 0.303
0.278 0.293
0.296 0.302
0.301 0.300
0.302 0.308
0.237 0.250
;
run;
proc means n mean std clm t prt;
  var diff;
run;
```

Figure 4.10. A confidence interval and test of hypothesis about μ_D

```
Analysis Variable : DIFF


   N         Mean        Std Dev   Lower 95.0% CLM  Upper 95.0% CLM
  ------------------------------------------------------------------
  10      0.0053000     0.0054171      0.0014249         0.0091751
  ------------------------------------------------------------------


                                    T    Prob>|T|
                               ------------------------
                               3.0939474   0.0128
                               ------------------------
```

being formed by subtracting the reading on the control member of the pair from the reading under the oxygen treatment, while in Rao's Example 4.11, the differences are computed in the other order. The test of

$$H_0 : \mu_D = 0$$

also could have been performed using the UNIVARIATE procedure on the variable DIFF. The *t*-value and (two-sided) *p*-value would be produced as standard output.

If it is desired to test

$$H_0 : \mu_D = \Delta_0$$

where Δ_0 is some value other than zero, then the MEANS or UNIVARIATE procedure could be performed using

$$dev = diff - \Delta_0$$

as the analysis variable.

4.6 Inferences about the Variance of a Single Population

Example 4.12. A laboratory manager wants to make sure that 95% of the carbon analysis measurements made in the laboratory are within 0.1 ppm of the true value. On the basis of past experience, the manager is willing to assume that the carbon content measurements made on identical soil samples are (approximately) normally distributed with mean μ and standard deviation σ. Recall that 95% of the measurements in a normal population will lie within 1.96 ($\approx$ 2) standard deviations of the mean. Therefore, the laboratory manager's requirement will be met by the measurements in the lab if $2\sigma < 0.10$ (that is, $\sigma < 0.05$).

A technician presented with ten identical soil samples for carbon analysis produced the following results (ppm) . . . Do these data indicate that the technician's measurements meet the lab manager's requirements? In other words, can the observed sample be regarded as a random sample from a population with standard deviation less than 0.05? To answer this question the manager can perform a hypothesis test in which the null and alternative hypothses are, respectively, H_0: $\sigma \geq 0.05$ and H_1: $\sigma < 0.05$.

Example 4.13. Now suppose that the manager in Example 4.12 wants to send the technician for further training if the standard deviation for the population of all measurements by this technician exceeds 0.05. On the basis of the given data, what would be your recommendation to the lab manager?

Rao's Box 4.6 shows how to compute a one- or two-sided confidence interval for a population variance and the test statistic for testing an hypothesis about the variance of a normal population. The confidence limits and test statistic depend on the sample size and the sample variance, and percentage points of the Chi-Square distribution. The sample size and variance can be obtained from SAS via the MEANS or the UNIVARIATE procedure, and the Chi-Square percentiles from the CINV command, as described in Chapter 2. From there, you are on your own for computing confidence limits and the value of the test statistic. In many instances, once you have the sample variance, it is probably most efficient for you to use a pocket calculator. However, the program in Figure 4.11 will test the hypotheses in Examples 4.12 and 4.13 and will compute the interval required in Example 4.13.

Let us look at the program in Figure 4.11. After the data are entered, the required sample information is obtained using the MEANS procedure. These statistics are outputted to a data set called "stats". This data set is shown in Figure 4.12. Then, in another data set called just "b", which uses the information in the data set "stats", first a Chi-Square test statistic is computed for testing

$$H_0 : \sigma = 0.05.$$

As seen in Figure 4.13, which is a printout of the data set "b", this statistic has value TSTSTAT = 22.4588 (which differs from Rao's value in Example 4.13 due just to rounding error).

Figure 4.11. SAS program to test an hypothesis about and compute confidence limits for a population variance

```
options ls=70 ps=54 pageno=1;
data a;
  input ppm @@;
lines;
0.560 0.842 0.731 0.782 0.673 0.718 0.791
0.726 0.760 0.798
;
run;
proc means n std var;
  var ppm;
  output out = stats n = n std=s var=var;
run;
proc print data = stats;
run;
data b; set stats;
  tststat = (n - 1)*var/(0.05)**2;
  xl1 = cinv(0.05,n-1);
  xl2 = cinv(0.025,n-1);
  xu1 = cinv(0.95,n-1);
  xu2 = cinv(0.975,n-1);
  ll1 = (n - 1)*var/xu1;
  ul1 = (n - 1)*var/xl1;
  ll2 = (n - 1)*var/xu2;
  ul2 = (n - 1)*var/xl2;
  lcl1 = ll1**0.5;
  ucl1 = ul1**0.5;
  lcl2 = ll2**0.5;
  ucl2 = ul2**0.5;
run;
proc print data = b;
run;
```

Figure 4.12. Statistics outputted from the MEANS procedure

OBS	_TYPE_	_FREQ_	N	S	VAR
1	0	10	10	0.078984	.0062385

The next four lines of the program in Figure 4.11 look up Chi-Square percentage points for one- and two-sided 95% confidence intervals. Keep in mind that while Rao is using notation corresponding to upper-tail

Figure 4.13. Computation of test statistics and confidence limits for a population variance

OBS	_TYPE_	_FREQ_	N	S	VAR	TSTSTAT	XL1
1	0	10	10	0.078984	.0062385	22.4588	3.32511

OBS	XL2	XU1	XU2	LL1	UL1	LL2
1	2.70039	16.9190	19.0228	.0033186	0.016886	.0029516

OBS	UL2	LCL1	UCL1	LCL2	UCL2
1	0.020792	0.057607	0.12995	0.054328	0.14419

Chi-Square tables, SAS's tables are cumulative, or lower-tail tables. Thus, what Rao would denote as

$$\chi^2\,(n-1,\,\alpha)$$

designating the point that cuts off $100\alpha\%$ of the area under the upper tail of the distribution, would be denoted by SAS as

$$\chi^2\,(n-1,\,1-\alpha).$$

The notation used in the program in Figure 4.11 is "xu1" stands for the Chi-Square value used to obtain the *u*pper limit for a *1*-sided interval, "xl2" is the Chi-Square value used in the *l*ower limit for a *2*-sided interval, etc. These Chi-Square values can also be used as the critical values for a one- or two-tailed $\alpha = 0.05$-level test of $H_0 : \sigma = \sigma_0^2$.

The lower and upper one- and two-sided confidence limits for the population variance are designated as ll1 ("lower limit, one-sided"), ll2, ul1, and ul2 in Figures 4.11 and 4.13, and the limits for the standard deviation are lcl1 ("lower confidence limit, one-sided"), lcl2, ucl1, and ucl2. Compare these results to those given in Rao's Example 4.13.

The program in Figure 4.11 can be used to compute test statistics, critical values, and confidence limits for σ (assuming $\alpha = 0.05$ is required) for other data sets by simply replacing the data of Rao's Example 4.12 with the appropriate other data.

4.7 Comparing Variances of Two Populations

Example 4.15. For the N-loss data in Example 3.27, assume normal populations and check whether it is reasonable to assume equal population variances, as *t*-test procedures require.

Recall that the TTEST procedure provides two versions of the two-sample *t*-statistic: one for use if the assumption of equal population variances can be made, and the other based on the assumption that the two populations are not equally variable. Also as part of the TTEST procedure, SAS provides a (two-sided) test of equality of variances of two normal populations,

$$H_0: \sigma_1^2 = \sigma_2^2 .$$

Figure 4.14 shows the TTEST procedure using the N-loss data from Figure 4.1.

Figure 4.14. TTEST analysis for N-loss data

```
                          TTEST PROCEDURE

Variable: NLOSS

FERT       N              Mean            Std Dev           Std Error
------------------------------------------------------------------------
u          13        8.58461538        2.28832040         0.63466589
un         15       11.14666667        2.13971516         0.55247208

Variances          T        DF      Prob>|T|
----------------------------------------------
Unequal      -3.0448      24.8       0.0054
Equal        -3.0600      26.0       0.0051

For H0: Variances are equal,  F' = 1.14     DF = (12,14)
                              Prob>F' = 0.8014
```

The last two lines of the TTEST procedure, beginning with the label,

```
For H0: Variances are equal
```

provide a test of whether the difference in the sample variances is more than can be attributed to chance, alone. The alternative hypothesis is the two-sided one, that the population variances are not equal.

In order to test this hypothesis, an F-statistic is formed as the ratio of the larger sample variance to the smaller. This ratio is denoted as F', and the larger the value of F', the stronger the evidence against H_0. Numerator and denominator degrees of freedom for the F'-statistic are based on the sizes of the samples with the larger and smaller sample variances, respectively. The probability of observing a ratio of variances this large purely due to sampling variability is given in Figure 4.14 as

```
Prob>F' = 0.8014.
```

In this example, then, there is little evidence that the populations have different amounts of variability, so that the appropriate test for difference in the population means is the pooled t-test. (Note that even though this F-statistic is formed so that only *large* values constitute evidence against H_0, this is still a two-sided test and the p-value is appropriate to the two-sided alternative hypothesis. If there had been some reason to hypothesize which population had the greater variability—as in Rao's Example 4.14 where the technician was given additional training in order to reduce the variability among measurements—then take half of the printed p-value.

SAS does not provide a built-in procedure for constructing a confidence interval for the ratio of two population variances. The MEANS procedure can be used to obtain sample variances, and the appropriate percentiles from the F-distribution can be obtained using the FINV function as described in Section 2.3, and then the desired interval can be constructed by hand. Neither does SAS compute prediction intervals for the mean of a future sample of m observations.

5

Ordinal Data

5.1 Introduction

The UNIVARIATE procedure performs the sign test and the signed ranks test for hypotheses concerning the location parameter of a single population. The rank sum test for comparing location parameters of two populations is available in the NPAR1WAY procedure.

5.2 Inferences Concerning the Location of a Single Population

Example 5.1. Schrier and Junga (1981) studied the entry and distribution of the drugs chlorpromazine (CPZ) and vinblastine (VBL) into human red blood cells (RBC) during endocytosis (process of absorption by plant and animal cells). The following data are the percentages of CPZ found in membrane fractions of red blood cells suspended in Hanks' solution or sucrose...

However, the procedures based on normally distributed populations may not be appropriate for these data. Practical experience indicates that instrument errors associated with the detection of low levels may cause the distribution of such measurements to be skewed to the right.

Example 5.2. Let's analyze the data in Example 5.1 to estimate the median percentage of CPZ in a membrane fraction of red blood cells suspended in Hanks' solution.

The small sample procedure outlined in Box 5.1 for constructing a confidence interval for the median of a population requires availability of tables of the binomial distribution. The program in Figure 5.1 produces the table of binomial probabilities for $n = 8$ and $\pi = 0.50$ in Figure 5.2.

Figure 5.1. SAS program for binomial probabilities

```
options ls=70;
data a;
  do i = 0 to 8;
  if i - 1 = -1 then p = probbnml(0.50,8,0);
  else p = probbnml(0.50,8,i) -
      probbnml(0.50,8,i-1);
  output;
  end;
run;
proc print data=a;
  title1 'Binomial Probabilities';
  title2 'for n = 8 and pi = 0.50';
run;
```

Figure 5.2. Binomial probability distribution for $n = 8$ and $\pi = 0.50$

```
                Binomial Probabilities
                for n = 8 and pi = 0.50

           OBS      I        P

            1       0      0.00391
            2       1      0.03125
            3       2      0.10938
            4       3      0.21875
            5       4      0.27344
            6       5      0.21875
            7       6      0.10938
            8       7      0.03125
            9       8      0.00391
```

In Figure 5.2, $p = \Pr[Y \le 0] = 0.00391$ and $\Pr[Y \le 1] = $).00391 + 0.03125 = 0.03516. If we take $\alpha/2 \approx 0.0352$, then since $\alpha = 0.0704$, a 92.96% confidence interval on the median can be constructed by taking the upper limit to be the 7th data value, in ascending order of magnitude, since $b(8, 0.0352) = 7$, and the lower limit to be the 2nd data value, since $n + 1 - 7 = 2$. Since the ordered sample consists of the values

5, 6, 7, 14, 14, 15, 17, 27,

the point estimate of the population median is the sample median

$$(14 + 14)/2 = 14,$$

and we are about 93% confident that the population median is between 6 and 17.

Example 5.3. Do the data in Example 5.1 indicate a difference in the percentage of CPZ in membrane fractions of red blood cells suspended in the two solutions?

The UNIVARIATE procedure may be used to perform the sign test of the null hypothesis that the population median of the paired differences is zero. The program is given in Figure 5.3, and the output in Figure 5.4, Parts 1 and 2.

Figure 5.3. Program to produce the sign and signed ranks tests

```
options ls=70 ps=54 pageno=1;
data a;
   input hanks sucrose;
   diff=hanks-sucrose;
lines;
  5   9
  6   4
  7   7
 15  13
 17  16
 14  22
 14  21
 27  20
 ;
run;
proc univariate plot;
   var diff;
run;
```

Figure 5.4, Part 1. UNIVARIATE output: statistics

```
                      Univariate Procedure

Variable=DIFF


                            Moments

        N                      8  Sum Wgts             8
        Mean              -0.875  Sum                 -7
        Std Dev         5.083236  Variance       25.83929
        Skewness        -0.11434  Kurtosis       -0.75314
        USS                  187  CSS             180.875
        CV              -580.941  Std Mean       1.797195
        T:Mean=0        -0.48687  Pr>|T|           0.6412
        Num ^= 0               7  Num > 0               4
        M(Sign)              0.5  Pr>=|M|          1.0000
        Sgn Rank            -2.5  Pr>=|S|          0.6875



                        Quantiles(Def=5)

        100% Max            7        99%             7
         75% Q3             2        95%             7
         50% Med          0.5        90%             7
         25% Q1          -5.5        10%            -8
          0% Min           -8         5%            -8
                                      1%            -8

        Range              15
        Q3-Q1             7.5
        Mode                2



                            Extremes

        Lowest     Obs       Highest    Obs
            -8(      6)           0(      3)
            -7(      7)           1(      5)
            -4(      1)           2(      2)
             0(      3)           2(      4)
             1(      5)           7(      8)
```

Although it is difficult to tell anything about the population distribution by looking at a sample of size $n = 8$, the plots in Part 2 of Figure 5.4 indicate the possibility of nonnormal distribution of differences. If it is desired to test the null hypothesis that the median difference in the population is zero using the sign test, look at the third-to-last line in the section labeled "Moments" in Part 1 of Figure 5.4:

Figure 5.4, Part 2. UNIVARIATE output: plots

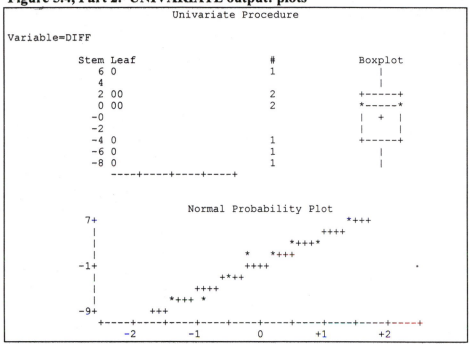

```
                         Univariate Procedure

Variable=DIFF

            Stem Leaf                      #           Boxplot
               6 0                         1              |
               4                                          |
               2 00                        2           +-----+
               0 00                        2           *-----*
              -0                                       |  +  |
              -2                                       |     |
              -4 0                         1           +-----+
              -6 0                         1              |
              -8 0                         1              |
                 ----+----+----+----+


                        Normal Probability Plot
            7+                                       *+++
             |                                    ++++
             |                               *++++*
             |                          *    *+++
           -1+                          ++++                        .
             |                       +*++
             |                   ++++
             |               *+++  *
           -9+          +++
             +----+----+----+----+----+----+----+----+----+----+
                -2        -1         0        +1        +2
```

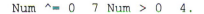

Num ^= 0   7 Num > 0   4.

In the set of eight differences, there are seven which are nonzero and four which are positive. The next-to-last line in the output is

M(Sign)   0.5   Pr>=|M| 1.000.

The test statistic SAS uses for the sign test is

$$M(\text{Sign}) = B_c - n/2,$$

where n is the number of nonzero differences in the sample, and B_c is the number of positive differences. Since in this example, there are $n = 7$ nonzero differences, of which $B_c = 4$ are positive:

$$M(Sign) = 4 - 7/2$$
$$= 4 - 3.5$$
$$= 0.5.$$

The table of binomial probabilities for $n = 7$ nonzero differences is given in Figure 5.5. It can be used to see where SAS obtains the p-value,

$$Pr>=|M| = 1.0000$$

and/or to construct a confidence interval for the median difference. The (one-sided) probability of obtaining 4 or more positive differences by chance alone (i. e., when $\pi = 0.5$) is

$$0.27344 + 0.16406 + 0.05469 + 0.00781 = 0.5000,$$

so that the probability reported by SAS for a two-tailed test is $2(0.5) = 1.00$.

Figure 5.5. Binomial distribution for $n = 7$ and $\pi = 0.50$.

```
            Binomial Probabilities
            for n = 7 and pi = 0.50

       OBS      I        P

        1       0      0.00781
        2       1      0.05469
        3       2      0.16406
        4       3      0.27344
        5       4      0.27344
        6       5      0.16406
        7       6      0.05469
        8       7      0.00781
```

Note that the sign test could be used to test the null hypothesis that the median of a single population, whether it be a population of differences or of actual measurements, is some specified value other than zero. As with the t-test, simply perform the UNIVARIATE procedure on data expressed as deviations from the hypothesized value.

Example 5.4. We now use the Wilcoxon signed ranks procedure to analyze the CPZ data in Example 5.3. In addition to the assumption that the population of differential CPZ values has a continuous distribution, we need to assume that the population is symmetric.

The signed ranks test of the null hypothesis that the median of the population of differences is zero is given in the last line of the section headed "Moments" in Part 1 of Figure 5.4:

```
Sgn Rank   -4        Pr>=|S|   0.5625.
```

The version of the signed ranks statistic reported by SAS is the so-called Mann-Whitney form, defined as

$$\text{Sgn Rank} = W^+ - \frac{N}{2}$$

where

$$W^+ = \text{sum of the ranks with positive signs}$$

and

$$N = \frac{n(n+1)}{2}$$

Under H_0 you would expect that this total be split equally between ranks with positive signs and those with negative signs. In this example, the ranks of the seven nonzero differences would total $7(8)/2 = 28$ and we would expect under H_0 that the positive ranks would total half of this, or 14. In fact, our sample gives the total of the positive ranks to be 10 which is four fewer than expected:

$$\text{Sgn Rank} = 10 - 14 = -4.$$

The (two-sided) probability of an occurrence this different from what is expected by chance alone is given as $p = 0.5625$.

In order to use the signed ranks test to test whether the median of a population, be it a population of actual measurements or a population of differences, is some value other than zero, simply perform the UNIVARIATE procedure on data expressed as deviations from the hypothesized value.

In the Appendix to this chapter is a SAS program for computing the Walsh averages shown in Rao's Example 5.4.

5.3 Inferences Concerning the Difference in Location of Two Populations

Example 5.6. During an *in vivo* comparison of two antitumor drugs conducted in the laboratory of Dr. Norman Reed, Department of Microbiology, Montana State University, Collings and Hamilton (1988) collected the following data on negative (natural) log-transformed tumor weights for two groups of 12 mice, each exposed to a different drug.

Collings and Hamilton note that the negative log-transformed data conform relatively well to the two-sample shift model in Box 5.2, and that both distributions appear to be positively skewed (i.e., the right tail is longer). Thus, we should look for an alternative to the *t*-test procedure for inferences about the shift between the two population distributions (which is the same as the difference between medians)

The SAS procedure NPAR1WAY contains a variety of "nonparametric" (NPAR) procedures for comparing location parameters for "one-way" studies. A synonym for nonparametric is "distribution free", meaning that the significance probabilities, critical values, etc., do not depend on the distribution of the population. For our present purposes, "one-way" refers to the fact that there is only one variable listed in the CLASS statement denoting the variable that defines the groups to be compared.

Figure 5.6 shows the program for analyzing these data using the Wilcoxon rank sum test. Since NPAR1WAY can produce quite a variety of analyses, it is necessary to specify the Wilcoxon test,

```
proc nparlway wilcoxon;
```

Figure 5.6. SAS program to produce Wilcoxon rank sum test

```
options ls=70 ps=54 pageno=1;
/*Wilcoxon Rank Sum Test*/
data a;
  input group$ @@;
    do i = 1 to 12;
      input neglnwt @@;
      output;
    end;
  drop i;
lines;
std -1.56444 -1.52562 -1.33500 -1.11711
        -1.05222 -0.94507 -0.80648 -0.54349
        -0.01489 0.16487 0.46681 1.80181
test -0.95089 -0.94663 -0.69315 -0.59222
        -0.55962 -0.52117 -0.00995 0.07472
        0.14156 0.19601 0.65887 2.93746
;
run;
proc sort;
  by group;
run;
proc univariate plot;
  var neglnwt;
  by group;
run;
proc npar1way wilcoxon;
  class group;
  var neglnwt;
run;
```

in order to limit the output to just the rank sum test. Note that the commands are quite similar to those for the two-sample *t*-test: a CLASS statement names the variable that defines the two populations, and the VAR statement specifies the variable whose medians we wish to compare. Furthermore, the data in the two groups are independent measurements, just as was the case for the two-sample *t*-test. Additionally, side-by-side box-and-whisker plots are obtained from the UNIVARIATE procedure to supplement the analysis by the Wilcoxon procedure. The output from NPAR1WAY and the side-by-side box-and-whisker plots (only) from UNIVARIATE are shown in Figure 5.7.

Figure 5.7. Side-by-side box-and-whisker plots for tumor data

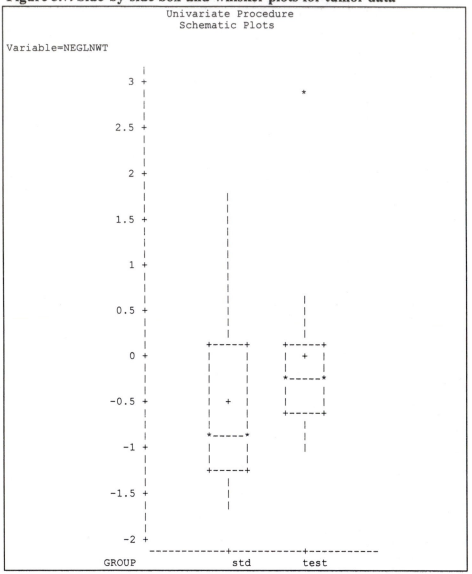

The box-and-whisker plots indicate rather ill-behaved data, so that the rank sum test is more appropriate than the two-sample *t*-test. Figure 5.8 gives the *large-sample* version of the Wilcoxon test, which is marginally appropriate since both samples are of size $n = 12$. Ranks 1 through 24 are assigned to the combined data. The ranks 1 through 24 sum to $24(25)/2 = 300$. The observed rank sums for the standard and test drugs are, respectively, 123 and 177, so that the average ranking of animals receiving the standard drug is 10.25, while the mean ranking of animals receiving the test drug is 14.75. If the null hypothesis of no difference in population medians is true, then we would expect the sum of the ranks from each of the two samples to be (apart from sampling error) one-half of the total, or 150. The standard deviations of the expected ranks in the two samples is given as 17.32. Do the observed rank sums differ too much from the expected to attribute solely to chance?

Figure 5.8. Wilcoxon rank sum test

```
               N P A R 1 W A Y   P R O C E D U R E

         Wilcoxon Scores (Rank Sums) for Variable NEGLNWT
                  Classified by Variable GROUP


                          Sum of     Expected     Std Dev        Mean
      GROUP       N       Scores     Under H0     Under H0       Score

      std         12      123.0       150.0      17.3205081   10.2500000
      test        12      177.0       150.0      17.3205081   14.7500000

        Wilcoxon 2-Sample Test (Normal Approximation)
        (with Continuity Correction of .5)

        S=  123.000    Z= -1.52998    Prob > |Z| =   0.1260

        T-Test approx. Significance =      0.1397

        Kruskal-Wallis Test (Chi-Square Approximation)
        CHISQ=  2.4300   DF=  1   Prob > CHISQ=      0.1190
```

SAS chooses the sum of the ranks in the smaller sample, if sample sizes are unequal, or the sum of the ranks in the first (alphabetically) group if the sample sizes are equal, and computes the corresponding standard score:

$$z = \frac{123 - 150 + .5}{17.3205} = -1.52998.$$

Since the set of ranks are discrete but the normal distribution applies to continuous data, a "continuity correction" of 0.50 is applied to get a more accurate probability. If the sum of the ranks is less than expected, then 0.5 is added in the numerator; otherwise, 0.5 is subtracted in the numerator of the z-statistic.

The value $z = -1.52998$ indicates that the observed sum of ranks is 1.5 standard deviations below what was expected under the null hypothesis. The (two-sided) probability of a deviation this large, in either direction, by chance alone is given as

```
Prob > |z| = 0.1260.
```

If the value $z = -1.52998$ were referred to tables of the t-distribution with $11 + 11 = 22$ degrees of freedom, the (two-sided) p-value would be approximately $p = 0.1397$.

The WILCOXON option can be used for performing the Kruskal-Wallis test described in Section 8.4 (Rao's Section 8.8) for comparing location parameters for more than two populations, and when this is done, no z-statistics are printed but the appropriate test of equality of all population locations is the Chi-Square test. When there are only two populations as in Figure 5.8, the Kruskal-Wallis Chi-Square equals the square of the z for the Wilcoxon test (without continuity correction):

```
CHISQ = 2.43.
```

Example 5.7. In a study of serum albumin levels in children with clinical evidence of protein-energy malnutrition (PEM), the following serum albumin levels (g/dl) were obtained for four control children and three children with a form of PEM known as marasmus.

Let's use these data to verify the research hypothesis that the median serum albumin level for control children is 1.5 g/dl more than the median level for children with the marasmus form of PEM.

The NPAR1WAY procedure in SAS should not be used to analyze data from such small samples as those described in Example 5.7. However, it is possible to use the NPAR1WAY procedure to test an hypothesis of the form

$$H_0: \mu_1 - \mu_2 = \Delta,$$

where Δ is a value other than zero, by simply expressing the observations in the sample hypothesized to have the larger median as deviations from the value Δ.

5.A. Appendix: A SAS Program to Compute Walsh Averages

In his Section 5.4, Rao defines Walsh averages and illustrates how to compute them and use them to test hypotheses concerning the population median and construct confidence intervals on the population median. The SAS program in Figure 5.9 computes the Walsh averages for the CPZ data of Rao's Example 5.1.

Use of arrays in Figure 5.9 is not explained in this text. Refer to Delwiche and Slaughter, Cody and Smith, or *SAS Language and Procedures* for explanations of arrays. Otherwise, from the printout of data set "b" in Figure 5.10, you can probably follow what the programming steps do.

To use this program on another set of data, the only changes that need to be made are: (1) to supply appropriate data and variable names and (2) to change the 8 to whatever n is in the new data set in the two lines, `array` and `retain`.

Figure 5.9. SAS program to compute Walsh averages for CPZ data

```
options ls = 70 ps = 54 pageno = 1;
/*Walsh Averages*/
data a;
  input hanks sucrose;
  diff = hanks - sucrose;
lines;
 5   9
 6   4
 7   7
15  13
17  16
14  22
14  21
27  20
;
run;
proc sort data = a;
  by diff;
run;
data b; set a;
  case = _n_;
  array d{8} d1 - d8;
  retain d1 - d8;
  d{case} = diff;
  do j = 1 to case;
    walsh = (d{j} + d{case})/2;
    output;
  end;
run;
proc print data = b;
run;
proc sort data = b;
  by walsh;
run;
proc print data = b;
  var walsh;
run;
```

Figure 5.10. Output from program to compute Walsh averages

OBS	HANKS	SUCROSE	DIFF	CASE	D1	D2	D3	D4	D5	D6	D7	D8	J	WALSH
1	14	22	-8	1	-8	.	.	.	.	.	.	.	1	-8.0
2	14	21	-7	2	-8	-7	.	.	.	.	.	.	1	-7.5
3	14	21	-7	2	-8	-7	.	.	.	.	.	.	2	-7.0
4	5	9	-4	3	-8	-7	-4	.	.	.	.	.	1	-6.0
5	5	9	-4	3	-8	-7	-4	.	.	.	.	.	2	-5.5
6	5	9	-4	3	-8	-7	-4	.	.	.	.	.	3	-4.0
7	7	7	0	4	-8	-7	-4	0	.	.	.	.	1	-4.0
8	7	7	0	4	-8	-7	-4	0	.	.	.	.	2	-3.5
9	7	7	0	4	-8	-7	-4	0	.	.	.	.	3	-2.0
10	7	7	0	4	-8	-7	-4	0	.	.	.	.	4	0.0
11	17	16	1	5	-8	-7	-4	0	1	.	.	.	1	-3.5
12	17	16	1	5	-8	-7	-4	0	1	.	.	.	2	-3.0
13	17	16	1	5	-8	-7	-4	0	1	.	.	.	3	-1.5
14	17	16	1	5	-8	-7	-4	0	1	.	.	.	4	0.5
15	17	16	1	5	-8	-7	-4	0	1	.	.	.	5	1.0
16	6	4	2	6	-8	-7	-4	0	1	2	.	.	1	-3.0
17	6	4	2	6	-8	-7	-4	0	1	2	.	.	2	-2.5
18	6	4	2	6	-8	-7	-4	0	1	2	.	.	3	-1.0
19	6	4	2	6	-8	-7	-4	0	1	2	.	.	4	1.0
20	6	4	2	6	-8	-7	-4	0	1	2	.	.	5	1.5
21	6	4	2	6	-8	-7	-4	0	1	2	.	.	6	2.0
22	15	13	2	7	-8	-7	-4	0	1	2	2	.	1	-3.0
23	15	13	2	7	-8	-7	-4	0	1	2	2	.	2	-2.5
24	15	13	2	7	-8	-7	-4	0	1	2	2	.	3	-1.0
25	15	13	2	7	-8	-7	-4	0	1	2	2	.	4	1.0
26	15	13	2	7	-8	-7	-4	0	1	2	2	.	5	1.5
27	15	13	2	7	-8	-7	-4	0	1	2	2	.	6	2.0
28	15	13	2	7	-8	-7	-4	0	1	2	2	.	7	2.0
29	27	20	7	8	-8	-7	-4	0	1	2	2	7	1	-0.5
30	27	20	7	8	-8	-7	-4	0	1	2	2	7	2	0.0
31	27	20	7	8	-8	-7	-4	0	1	2	2	7	3	1.5
32	27	20	7	8	-8	-7	-4	0	1	2	2	7	4	3.5
33	27	20	7	8	-8	-7	-4	0	1	2	2	7	5	4.0
34	27	20	7	8	-8	-7	-4	0	1	2	2	7	6	4.5
35	27	20	7	8	-8	-7	-4	0	1	2	2	7	7	4.5
36	27	20	7	8	-8	-7	-4	0	1	2	2	7	8	7.0

Figure 5.11. Walsh averages in order of magnitude

OBS	WALSH
1	-8.0
2	-7.5
3	-7.0
4	-6.0
5	-5.5
6	-4.0
7	-4.0
8	-3.5
9	-3.5
10	-3.0
11	-3.0
12	-3.0
13	-2.5
14	-2.5
15	-2.0
16	-1.5
17	-1.0
18	-1.0
19	-0.5
20	0.0
21	0.0
22	0.5
23	1.0
24	1.0
25	1.0
26	1.5
27	1.5
28	1.5
29	2.0
30	2.0
31	2.0
32	3.5
33	4.0
34	4.5
35	4.5
36	7.0

6

Categorical Data

6.1 Introduction

SAS can be used to compute p-values for normal, binomial, and Chi-Square tests concerning proportions in a single population and for the McNemar (sign) test to compare proportions in two related populations. The FREQuency procedure can be used to perform the Chi-Square test of independence and Fisher's exact test comparing proportions in two or more populations.

6.2 Inferences About Population Proportions

Example 6.1. To assess the effectiveness of a new treatment for aphid infestation of tobacco plants, 100 treated plants were classified into three infestation categories. It was found that 67 plants had no aphid infestation, 24 had only stem infestation and 9 had only leaf infestation.

Example 6.3. Suppose that in Example 6.1, we want to determine whether the observed data justify the claim that less than 15% of the population of treated plants will have only leaf infestation.

Let π denote the population proportion of plants with only leaf infestation. Then π is estimated from these data by $\hat{\pi} = 9/100 = 0.09$. Since this is a large sample, the null hypothesis $H_0 : \pi = 0.15$ can be tested against the research hypothesis $H_1 : \pi < 0.15$ using a z-test. Rao computes the value of the z-statistic to be

$$z = -2.10$$

in Example 6.3. The PROBNORM function in SAS can be used to compute the probability of obtaining a z-value this small or smaller by chance alone.

Recall that the PROBNORM function gives cumulative normal probabilities; that is,

```
probnorm(x)
```

is the probability that a normally distributed random variable will have a value less than or equal to x. If it is desired to compute the probability that the value is *greater* than x, then subtract the PROBNORM value from 1. Figure 6.1 shows the SAS code for computing the p-value for the test described in Example 6.3. Note that the function is called up in a data step, and then the resulting data set is printed. Figure 6.2 shows the resulting value.

Figure 6.1. Using the PROBNORM function to compute a p-value

```
data a;
   p=probnorm(-2.10);
run;
proc print data=a;
   title 'p-Value for z-test';
run;
```

Figure 6.2. p-value for the z-test

```
               p-Value for z-test

          OBS           P

           1       0.017864
```

In order to construct a confidence interval on π, once a confidence level is specified, the SAS function PROBIT can be used to look up the z-value which corresponds to the desired confidence level. For example, in case you have forgotten that $z = 1.96$ for a two-sided 95% confidence interval, writing

$$z = \text{probit}(0.975);$$

will return the value $z = 1.96$. Writing

$$z = \text{probit}(0.025);$$

will return the value $z = -1.96$, since SAS uses the *cumulative* normal distribution. The confidence limits can then be computed by hand using the formula in Rao's Box 6.1.

Example 6.4. A plant breeder observed two dwarf plants among $n = 10$ F_2 - generation plants. Do these data indicate that less than 25% of all F_2 - generation plants will be dwarfs?

Since the sample size in this example is quite small, what is desired is the probability that 2 or fewer dwarf plants out of 10 would be observed strictly by chance. The SAS function PROBBNML gives cumulative binomial probabilities.

Recall that the general form of the binomial probability function in SAS is

$$\text{probbnml}(\pi, n, r)$$

where π is the value specified in H_0, n is the sample size, and r is the observed number of "successes". Writing

$$p = \text{probnml}(0.25, 10, 2)$$

in a data step as in Figure 6.1 produces the probability shown in Figure 6.3.

Figure 6.3. *p*-value for the binomial test

```
                    p-Value for Binomial Test

                    OBS          P

                     1        0.52559
```

Example 6.8. Over a specified period, observers sighted 200 birds at a particular location. The birds may be classified into four species as follows:

Species	1	2	3	4	Total
Number of birds	40	80	65	15	200

Do these data provide evidence that the composition of the species in the location has changed from a previously known proportion of 3:3:3:1?

SAS does not provide the Chi-Square test of goodness of fit for discrete distributions. (The CAPABILITY procedure in SAS/QC, the programs used in quality control applications, can be used to perform the Chi-Square goodness of fit test to several continuous distributions commonly encountered in engineering applications.) However, the PROBCHI function can be used to find the *p*-value once the value of the Chi-Square statistic is computed by hand.

Rao finds

$$\chi^2 = 15$$

with 4 degrees of freedom in Example 6.8. Since large values for χ^2 indicate disagreement of the data with the proposed ratio, an upper-tail Chi-Square probability is required. As with the other distributions we have used so far, the probabilities calculated by this SAS function are cumulative (less-than or equal-to) probabilities; thus, what we require is

```
1 - probchi(x,df)
```

where the first value in the parentheses is the observed value for the test statistic and the second is degrees of freedom (15 and 3, respectively, in this example). Invoking this function in a program like the one in Figure 6.1 returns the value in Figure 6.4.

Figure 6.4. *p*-value for the Chi-Square test

```
                    p-Value for Chi-Square Test

                   OBS            P

                    1        .0018166
```

6.3 Comparing Proportions in Two Populations

Independent Samples

Example 6.10. In Example 6.1. we described the frequency distribution of types of aphid infestation of treated tobacco plants.   The observed frequency distribution of aphid infestation in 100 treated tobacco plants (from Example 6.1) and 100 untreated plants is as follows.. The following questions might be asked about these data.

1. Is the proportion of leaf infestation in untreated plants higher than that in treated plants? What is a 95% confidence interval for the difference between these two proportions?
2. Are the proportions of plants in the various infestation categories different in the populations of treated and untreated plants? In other words, are the population distributions across the three infestation categories the same for the treated and untreated plants?

Example 6.13. In Example 6.11 the category representing no infestation corresponded to $k = 3$. Let π_{13} and π_{23} denote, respectively, the proportion of uninfested plants in the treated and control populations... Suppose we want to scrutinize the claim that the proportion of treated plants that remain free of infestation is larger. Then, we should test the null hypothesis H_0: $\pi_{13} - \pi_{23} \leq 0$ against the research hypothesis H_1: $\pi_{13} - \pi_{23} > 0$.

In Example 6.13 Rao finds

$$z = 8.31.$$

Using

```
p = 1 - probnorm(8.31)
```

in a program like the one in Figure 6.1, since an upper-tail probability is required here, we find $p = 0$, to 6 decimal places. Confidence intervals on $\pi_1 - \pi_2$ can be formed as in Example 6.2.

> **Example 6.14.** A random sample of six sites in each of two North Florida counties is classified into three categories: predominantly pine, predominantly hardwood and other... Let π_1 and π_2 denote, respectively, the proportions of predominantly pine forest sites in the two counties. Fisher's exact test can be used to test the null hypothesis H_0: $\pi_1 \geq \pi_2$ against the alternative H_1: $\pi_1 < \pi_2$.

The FREQuency procedure in SAS performs Fisher's exact test. Figure 6.5 shows a program to analyze the data of Exercise 6.14. We introduced the FREQ procedure in Section 3.4 and used it to obtain a univariate frequency table. In the present example, we are using the FREQ procedure to construct a *bivariate* frequency distribution, also called a *two-way table,* a *contingency table,* or a *crosstabulation (crosstab).* The TABLES statement,

```
tables county*type;
```

instructs SAS to construct a two-way table with values of the variable COUNTY indicating rows and values of TYPE as column headings. Note that since the data have already been summarized into a two-way table, each cell in the table is a case. The INPUT statement identifies the value of the row variable (COUNTY), the value of the column variable (TYPE), and the cell frequency (SITES). Then the

```
weight sites;
```

command is used in the FREQ procedure to indicate cell counts. (Refer to Figure 3.6, in which a one-way frequency table was used as input to the CHART procedure, and to Section 6.4.)

The program in Figure 6.5 uses the FORMAT procedure and the FORMAT command in the FREQ procedure. In the interest of continuity, these will be explained in Section 6.5. Invoking the option EXACT in the TABLES statement causes a variety of statistics about the table to be printed, as is seen in the output in Figure 6.6.

Figure 6.5. Program to perform Fisher's exact test

```
options ls=70 ps=54 pageno=1;
/*Fisher's Exact Test*/
data a;
 input county type sites;
lines;
1 1 1
1 2 5
2 1 4
2 2 2
;
run;
proc format;
   value type 1 ='Pine' 2 = 'Not Pine';
proc freq data=a;
   tables county*type/exact;
   weight sites;
   format type type.;
run;
```

The first three statistics in Figure 6.6 are various forms of the Chi-Square statistic. Note, however, the warning printed at the bottom of the page that the sample size is too small to use the Chi-Square test. The Mantel-Haenszel Chi-Square is not pertinent to this example: it is used when both the row and column headings can be ordered from low to high and when testing for increasing or decreasing relationships.)

Figure 6.6. Fisher's exact test

```
                    TABLE OF COUNTY BY TYPE

          COUNTY       TYPE

          Frequency|
          Percent  |
          Row Pct  |
          Col Pct  |Pine    |Not Pine|  Total
          ---------+--------+--------+
                1 |     1 |     5 |     6
                  |  8.33 | 41.67 | 50.00
                  | 16.67 | 83.33 |
                  | 20.00 | 71.43 |
          ---------+--------+--------+
                2 |     4 |     2 |     6
                  | 33.33 | 16.67 | 50.00
                  | 66.67 | 33.33 |
                  | 80.00 | 28.57 |
          ---------+--------+--------+
          Total         5       7      12
                     41.67   58.33  100.00


            STATISTICS FOR TABLE OF COUNTY BY TYPE

          Statistic              DF    Value      Prob
          -------------------------------------------------
          Chi-Square              1    3.086      0.079
          Likelihood Ratio Chi-Square  1  3.256   0.071
          Continuity Adj. Chi-Square   1  1.371   0.242
          Mantel-Haenszel Chi-Square   1  2.829   0.093
          Fisher's Exact Test (Left)              0.121
                              (Right)             0.992
                              (2-Tail)            0.242
          Phi Coefficient             -0.507
          Contingency Coefficient      0.452
          Cramer's V                  -0.507

          Sample Size = 12
          WARNING: 100% of the cells have expected counts less
                   than 5. Chi-Square may not be a valid test.
```

The *p*-values for upper-, lower-, and two-tailed versions of Fisher's exact test follow in Figure 6.6. In the row labeled "(Left)" we find the *p*-value, $p = 0.121$, which Rao obtained in Example 6.14. Note that the order in which the rows and columns of the table are listed will affect whether a *p*-value is listed as a left or a right value, and recall that by default, SAS will construct tables which list values in either numerical order (if the variable is a numerical variable) or alphabetical order (if the variable is declared to be a character variable in the INPUT statement). Since the research hypothesis is that County 1 will have fewer sites classified as pine forest than County 2, an outcome more extreme than the observed data will have a *lower* frequency in the (1,1) cell. *SAS/STAT User's Guide, Volume 1*, page 866 explains, "For two-tailed tests, A [the set of tables whose probabilities are summed to get the *p*-value] is the set of tables with *p* less than or equal to the probability of the observed table. For left-tailed (right-tailed) tests, A is the set of tables where the frequency in the (1,1) cell is less than (greater than) or equal to that of the observed table."

Before leaving this example, we might mention that the three coefficients at the bottom of Figure 6.6--the phi coefficient, the contingency coefficient, and Cramer's *V*--are various types of correlation coefficients, measuring the degree of association in the table, or in this example, the extent to which the two counties differ. Values of zero for these coefficients indicate no difference in the two populations, and the farther away from zero, the greater the extent to which the two counties differ. If the populations have a natural ordering and so do the responses, then the signs on the phi and Cramer's coefficients have meaning; the contingency coefficient is always positive. Although in this example, the sizes of the coefficients (in absolute value) are roughly comparable, this is not always the case, because in general, the three coefficients do not all have an upper limit of 1 (in absolute value) or even the same upper limit. Rather, the maximum possible value depends on the dimension of the table, and unless you have at hand what the maximum value could be, the size of the coefficient is difficult to interpret. Unless you have made a study of measures of association in contingency tables and are familiar with the meaning of the sizes of the various coefficients, these measures of association are probably most useful when comparing two tables. For example, suppose that one lumber broker produced the table in Example 6.14 and another lumber broker produced slightly different classifications of the sites in the two counties. Comparing the coefficients would give a comparison of which broker found the greater difference in the two counties.

Example 6.16. Let us see if the data in Example 6.10 support the claim that the rates of types of aphid infestation are different for treated and untreated tobacco plants.

The Chi-Square test of independence, can be performed using the FREQ procedure, as we have seen. Figure 6.7 shows the program to analyze the data of Example 6.10. The output is given in Figure 6.8. The results of the Chi-Square test are as seen in Rao's Example 6.16.

Figure 6.7. Program to perform Chi-Square test of independence

```
options ls=70 ps=54 pageno=1;
/*Chi-Square Test of Independence*/
data a;
  input tmt infest plants;
lines;
1 1 9
1 2 24
1 3 67
2 1 39
2 2 44
2 3 17
;
run;
proc format;
  value t 1 ='Treated' 2 = 'Untreated';
  value i 1 = 'Leaf' 2 = 'Stem & Leaf' 3 =
      'None';
run;
proc freq;
  tables tmt*infest/chisq;
  weight plants;
  format tmt t.;
  format infest i.;
run;
```

Figure 6.8. Chi-Square test comparing treated and untreated tobacco

```
                    TABLE OF TMT BY INFEST

       TMT           INFEST

       Frequency |
       Percent   |
       Row Pct   |
       Col Pct   |Leaf     |Stem & L|None    |
                 |         |eaf     |        |   Total
       ----------+--------+--------+--------+
       Treated   |      9 |     24 |     67 |    100
                 |   4.50 |  12.00 |  33.50 |  50.00
                 |   9.00 |  24.00 |  67.00 |
                 |  18.75 |  35.29 |  79.76 |
       ----------+--------+--------+--------+
       Untreated |     39 |     44 |     17 |    100
                 |  19.50 |  22.00 |   8.50 |  50.00
                 |  39.00 |  44.00 |  17.00 |
                 |  81.25 |  64.71 |  20.24 |
       ----------+--------+--------+--------+
       Total           48       68       84       200
                    24.00    34.00    42.00    100.00


            STATISTICS FOR TABLE OF TMT BY INFEST

    Statistic                      DF     Value      Prob
    ------------------------------------------------------
    Chi-Square                      2     54.394     0.000
    Likelihood Ratio Chi-Square     2     58.014     0.000
    Mantel-Haenszel Chi-Square      1     50.733     0.000
    Phi Coefficient                        0.522
    Contingency Coefficient                0.462
    Cramer's V                             0.522

    Sample Size = 200
```

Paired Samples

Example 6.12. In a controlled clinical trial to determine the efficacy of an experimental drug for treating migraine headache, each of 90 patients was treated with two drugs--the experimental drug and a placebo--in random order. Each treatment lasted 12 weeks. At the end of the treatment period, the effect of the drug was classified into three categories: completely effective (CE); somewhat effective (SE); and not effective (NE).

Example 6.15. Let's see if the migraine headache data in Example 6.12 indicate that the proportion of patients for whom the treatment is at least somewhat effective--that is, whose response is SE or CE--is higher for the experimental drug than for the placebo.

In general terms, the McNemar test is appropriate when each observational unit is subjected to two conditions, which can be denoted as treatment (or "drug" or "after") and control (or "placebo" or "before"). Similarly, under each condition, one of two responses may be recorded, which can be called "success" and "failure". In Rao's Example 6.15, a "success" is "at least somewhat improved," denoted as Category 1 while a failure is "not at all improved" or not Category 1.

Figure 6.9 shows the program for performing the McNemar test in SAS's FREQ procedure. For brevity, the "success" category is denoted "1" and failure by "0" in the program. The AGREE option in the TABLES statement will produce the McNemar test whenever the size of the table is 2 × 2.

Figure 6.9. SAS program to perform McNemar test

```
options ls=70 ps=54;
data a;
   input drug placebo count;
lines;
1 1 40
1 0 33
0 1 10
0 0 7
;
run;
proc freq data = a;
   tables drug * placebo/agree;
   weight count;
run;
```

It is common to see the McNemar test statistic presented as a special case of the Chi-Square statistic, which is the *square* of the z-statistic which Rao presents. This is the form given by SAS, as you can see in Figure 6.10. Note that the square of Rao's z-value, $z = 3.51$, is the value of the McNemar statistic, 12.302. The two-sided *p*-value is $p < 0.001$ for either form of the test.

The *kappa* coefficient in Figure 6.10 was developed as a measure of *interrater agreement.* For example, suppose you have a standard, but expensive method for classifying subjects as to whether or not they have a certain psychological disorder. A new, cheaper method is proposed. A group of *n* subjects is classified as "success" (have the disorder) or "failure" by the standard procedure, and then again independently using the new procedure. The ratings are summarized into a 2×2 table similar to that in Figure 6.10.

If there is perfect agreement between the two classification systems, the kappa coefficient will have the value +1. When the degree of agreement is less than perfect but still exceeds chance, kappa is positive. The closer the value is to +1, the greater the degree of agreement.

Negative values indicate *less* agreement than would be expected by chance. The closer the value is to −1, the greater the extent to which the two raters disagree.

As is the case with any Chi-Square test, the McNemar test is a large-sample test. If a small-sample version is required, the easiest thing to do is to perform a sign test (refer back to Section 5.2 and Rao's Section 5.3). Consider the data as arranged in a 2×2 table, where, as in Rao's Box 6.7, the letters *A, B, C,* and *D* indicate cell frequencies, and each of the $A + B + C + D = n$ observational units is categorized into one of the four cells of the table.

	Control	
Treatment	**Success = 1**	**Failure = 0**
Success = 1	*A*	*B*
Failure = 0	*C*	*D*

Figure 6.10. McNemar test for Example 6.15

```
TABLE OF DRUG BY PLACEBO

                DRUG        PLACEBO

                Frequency|
                Percent  |
                Row Pct  |
                Col Pct  |       0|       1|   Total
                ---------|--------|--------|
                      0  |     7  |    10  |     17
                         |  7.78  | 11.11  |  18.89
                         | 41.18  | 58.82  |
                         | 17.50  | 20.00  |
                ---------|--------|--------|
                      1  |    33  |    40  |     73
                         | 36.67  | 44.44  |  81.11
                         | 45.21  | 54.79  |
                         | 82.50  | 80.00  |
                ---------|--------|--------|
                Total          40       50       90
                            44.44    55.56   100.00


             STATISTICS FOR TABLE OF DRUG BY PLACEBO

                          McNemar's Test
                          --------------
        Statistic = 12.302      DF = 1        Prob = 0.001


                    Simple Kappa Coefficient
                    ------------------------
                                       95% Confidence Bounds
        Kappa = -0.027    ASE = 0.088       -0.198     0.145


        Sample Size = 90
```

Note that if the treatment is successful in changing failures to successes, then *B* will be a large number, relative to *C*.

Write a SAS data step as follows:

```
data;
        input treat cntrl count;
        diff = treat - cntrl;
lines;
1 1 A
```

```
1 0 B
0 1 C
0 0 D
```

The values of the variable DIFF will be -1, 0, and 1, with frequencies C, $(A + D)$, and B, respectively. Applying the sign test, B_c = the number of positive differences = B, and the number of nonzero differences is $B + C$. If the version of the test statistic given by SAS,

$$M(\text{Sign}) = B_c - n/2 = B - \frac{B+C}{2}$$

gives a large positive value, this is consistent with the research hypothesis.

Figure 6.11 shows the SAS code for obtaining the McNemar/sign test for the data of Example 6.15. Note that the

```
freq count;
```

line of the program is used to indicate that the values of the variable COUNT are the cell frequencies. (Section 6.4 summarizes commands for using frequency tables as input data for various SAS procedures.)

Figure 6.11. SAS program to perform McNemar test as a sign test

```
options ls=70 ps=54;
data a;
  input drug placebo count;
  diff = drug - placebo;
lines;
1 1 40
1 0 33
0 1 10
0 0 7
;
run;
proc univariate data = a;
  var diff;
  freq count;
run;
```

In Figure 6.12, we find the value of the M(Sign) statistic,

```
M(Sign) = 11.5
```

and its associated *p*-value,

```
Pr>=|M|······0.0006.
```

line of the program is used to indicate that the values of the variable COUNT are the cell frequencies. (Section 6.4 summarizes commands for using frequency tables as input data for various SAS procedures.)

Figure 6.12. UNIVARIATE analysis for Example 6.15

```
                         Univariate Procedure

Variable=DIFF


                             Moments

            N                   90   Sum Wgts            90
            Mean          0.255556   Sum                 23
            Std Dev       0.645836   Variance      0.417104
            Skewness      -0.29701   Kurtosis      -0.67284
            USS                 43   CSS           37.12222
            CV           252.7183    Std Mean      0.068077
            T:Mean=0     3.753916    Pr>|T|          0.0003
            Num ^= 0            43   Num > 0             33
            M(Sign)          11.5    Pr>=|M|         0.0006
            Sgn Rank          253    Pr>=|S|         0.0002
```

6.4 Frequency Tables as Input to SAS Procedures

While SAS procedures are typically pretty uniform in regard to the way various commands work, one area in which there is surprising variation from one procedure to another is that of entering a one- or two-way frequency table as input data. In Section 6.3, we saw different commands in the FREQ and UNIVARIATE procedures, and in Chapter 3 yet another method was used with the CHART procedure. Since there is no logical way to figure out what method to use in which procedure, this section contains a summary for future reference.

Four elementary procedures that can accept frequency tables as input data are CHART, FREQ, MEANS and UNIVARIATE. Below are shown how to use frequency tables as input to each of these procedures. As an example, suppose you have a one-way frequency table showing the classification of two hundred birds into four species categories, as in Example 6.8.

The input statement and data are written as

```
input species birds;

1 40
2 80
3 65
4 15
```

To use the CHART procedure to construct a histogram with vertical bars, write

```
proc chart;
      vbar species/freq=birds;
```

To use the FREQuency procedure to reproduce the frequency table, write

```
proc freq;
      tables species;
      weight birds;
```

To obtain descriptive statistics, etc., using the MEANS procedure *if SPECIES were a numerical variable*, the commands are

```
proc means;
        var species;
        freq  birds;
```

and *if species were a numerical variable* to use the UNIVARIATE procedure, write

```
proc univariate;
        var species;
        freq birds;
```

(Note: While the *SAS Procedures Guide* says that you can indicate the frequency count by

```
weight birds;
```

in UNIVARIATE, doing so does not, in fact, give the same results as using

```
freq birds;)
```


6.5 Value Formats and Variable Labels

Value Formats

Note that in the program in Figure 6.5, both COUNTY and TYPE have been entered as numerical variables, even though, in fact, both variables are categorical. Using these data, SAS constructs a 2 × 2 table with row headings 1 and 2 and with column headings 1 and 2. Because tomorrow you probably won't remember whether Type 1 refers to "Pine" or "Not Pine" you would prefer to have the columns labeled "Pine" and "Not Pine," rather than "1" and "2." Of course, TYPE could have been entered as a character variable in the first place, but an alternative way to apply labels to values of a variable is to use *formats.*

The FORMAT procedure can be used anywhere in a SAS program, even in a data step. What it does is define a set of labels to be applied to a set of values, and gives the set of labels a name. In the program in Figure 6.5 the FORMAT procedure is invoked. Then follows the VALUE statement. The name TYPE in the value statement says that the set of labels is going to be referred to as the "type" labels. This word was chosen because these labels are going to be used to apply to values of the *variable* named type, although it is not necessary to give the same name to the labels as the variable on which they are going to be used. The labels are going to be that the value 1 means "Pine" and the value 2 means "Not Pine." If you look at your log file at this point in the program, it will give you a message that the format TYPE has been outputted (defined) and is now available for use. (You might also get a message that the format already exists. This is because SAS has some built-in formats, and you might have chosen to name your set of labels with the same name as is already applied to one of the built-in sets. This is usually not a problem, however.)

To have SAS apply the labels, you have to tell it to. This is done by the FORMAT statement in the procedure in which you want to use the labels, namely the FREQ procedure in this example. The general form of the FORMAT statement is

```
format varname labelname.;
```

where *varname* is the name of the variable and *labelname* is the name of the set of labels. *Note carefully* that the *labelname* is followed by a period before the semicolon. This is to keep SAS from being confused because the variable and the label both have the same name. The period is still required even if the labels do not have the same name as the variable to which they are to be applied.

Applying these labels causes the first column in the two-way table to be labeled "Pine," instead of "1" and the second column to be labeled "Not Pine," instead of "2." Had the county names been given, another set of labels could have been defined so that the rows of the table would have been labeled with names, instead of numbers.

The advantage to entering character data as if it were numerical and then applying formats, instead of entering character variables as such, is that you can control the *order* in which the values appear. Recall that SAS will always list numerical values in numerical order and character values in alphabetical order. You might not want the character values listed in alphabetical order, however. For example, suppose products are rated as "Superior," "Average" and "Inferior." You would want these values listed in that order in your table, instead of alphabetically. Using the number 1 to denote "Superior," 2 for "Average" and 3 for "Inferior" and then applying formats to these numerical values would produce the desired table.

Variable Labels

Formats are used to assign labels to *values* of a variable. Labels can also be assigned to *variable* names using the LABEL statement. The general form of a LABEL statement is

```
label var1='label1' var2 = 'label2' ... vark='labelk';
```

where "var1", "var2", ..., "vark" are the names of the *k* variables to which you want to assign labels, and "label1", "label2", ..., "labelk" are the respective labels. Labels can be up to 40 characters long, including blanks. The LABEL statement is used in a DATA step before the LINES line, and will apply to every reference to the variables to be labeled in subsequent programming steps.

Writing

```
label tmt='Treatment' infest = 'Level of Infestation';
```

in the data step in the program in Figure 6.7 would have resulted in the rows of the Chi-Square table in Figure 6.8 being labeled "Treatment" and the columns being labeled "Level of Infestation."

7

Designing Research Studies

7.1 Introduction

Random assignment of experimental units to treatment groups may be accomplished using the PLAN procedure. The SAS program FPOWTAB can be used to determine sample size required for testing hypotheses about a single population mean or the difference in the means of two populations.

7.2 Randomization in Experiments

Example 7.2. To compare the weight gains of chickens fed four experimental diets, 100 chickens were assigned to 20 pens; each pen received five birds. The pens were divided into four groups of five pens each and the four diets were assigned to the four groups at random. The birds in a pen were fed a common diet for 12 weeks. At the end of the study period, the gain in weight of each of the 100 birds was measured.

The randomization required in this study is done in two stages. First, the 100 chicks are randomly assigned to 20 pens. Then the 20 pens are randomly assigned to the four diets. The PLAN procedure can be used to accomplish this randomization. Figure 7.1 shows how to assign chicks to pens. Although the subsequent assignment of pens to diets is part of the *same* SAS program, that portion of the program is shown in Figure 7.5.

Figure 7.1. Randomization of chicks to pens using the PLAN procedure

```
options ls=70 ps=54 pageno=1;
data a;
  do chick=1 to 100;
    do j=1 to 20;
      if (5*j-4<=chick<=5*j) then pen=j;
    end;
    output;
  end;
  drop j;
run;
proc print data=a;
  title 'Unrandomized Assignment of Chicks to
      Pens';
run;
proc plan seed = 82844;
  factors chick=100;
  output data=a out=b;
  title1 'Random Assignment of 100 Chicks';
  title2 'to 20 Pens (Balanced)';
run;
proc print data=b;
run;
proc sort data=b;
  by chick;
run;
proc print data=b;
  title1 'Random Assignment of Chicks to
      Pens';
  title2 'In Order of Chick Number';
run;
```

Start with a systematic (or an arbitrary) assignment of chicks to pens, and store this unrandomized design in a SAS data set. Then call the PLAN procedure to scramble the assignment in a random fashion. The SAS data set denoted

```
data a;
```

assigns chicks numbered 1-5 to pen 1, chicks numbered 6-10 to pen 2, and so on until chicks 96-100 are assigned to pen 20. Presumably, each chick is identifiable by number, perhaps on a leg tag. Figure 7.2 shows a portion of the printout resulting from the PRINT command. Since the assignment in Figure 7.2 is not random, it is included here only for instructional purposes and typically would not be printed by the researcher.

Figure 7.2. A portion of the unrandomized assignment of chicks to pens

```
         Unrandomized Assignment of Chicks to Pens            1

             OBS      CHICK     PEN

              1         1        1
              2         2        1
              3         3        1
              4         4        1
              5         5        1
              6         6        2
              7         7        2
              8         8        2
              9         9        2
             10        10        2
             11        11        3
             12        12        3
             13        13        3
             14        14        3
             15        15        3
             16        16        4
             17        17        4
             18        18        4
             19        19        4
             20        20        4
```

The SAS code

```
proc plan;
factors chicks=100;
```

simply scrambles the chick numbers in a random fashion. The particular random scrambling is determined by the "seed" in Figure 7.1. Any number can be specified for the seed--your birth date, social security number, today's date, etc. If no seed is specified, SAS will use one based on the date and time. Figure 7.3 shows the random scrambling of chick numbers.

The random assignment of chicks to pens is shown in Figure 7.4. In the interest of brevity, again, only a portion of the output is shown. Sorting by chick number, as is done in the program in Figure 7.1, may give a listing of the assignment which is easier to use, but the corresponding printout is

Figure 7.3. Chick numbers scrambled using the PLAN procedure

```
                     Random Assignment of 100 Chicks            4
                        to 20 Pens (Balanced)

Procedure PLAN

Factor    Select  Levels   Order
------    ------  ------   -------
CHICK        100     100   Random


   CHICK
---+---+---+---+---+---+---+---+---+---+---+---+---+---+---+---+---+---+---+---+
    4   1   7  64  28  83  60  51  21  68  67  59  65  86  73  78  38
   62  27   6  42  22  89  94  46  33  13  40  56  50  57  36   2  35
   23  44  70  96  16  25  82  34  17  37  39  18  77  20  81  32  66
   63  91  12  85  53  92  52  31  29  30   3  55  49  71  95  69  80
   58  99  45  47  75 100  76  19  61  10  24  98  43  41   8  74  84
   15  79  54  87  11   9  26  72   5  90  93  97  14  48  88
```

Figure 7.4. A portion of the random assignment of chicks to pens

```
                     Random Assignment of 100 Chicks            5
                        to 20 Pens (Balanced)

                     OBS      CHICK      PEN

                       1         4        1
                       2         1        1
                       3         7        1
                       4        64        1
                       5        28        1
                       6        83        2
                       7        60        2
                       8        51        2
                       9        21        2
                      10        68        2
                      11        67        3
                      12        59        3
                      13        65        3
                      14        86        3
                      15        73        3
                      16        78        4
                      17        38        4
                      18        62        4
                      19        27        4
                      20         6        4
```

not shown here. The random assignment of chicks to pens is stored in a data set called "b" in the program in Figure 7.1.

In Figure 7.5 is shown the rest of the program: that portion which randomly assigns the 20 pens to the four diets. The SAS data set named "c" takes the pens with the chicks assigned as in data set "b" and systematically assigns the first five pens to Diet A, then next five to Diet B, the next five to Diet C, and the last five to Diet D. A portion of this systematic assignment is shown, again for instructional purposes only, in Figure 7.6. The PLAN procedure then randomly rearranges the pen numbers, as shown in Figure 7.7. This random assignment is stored in a SAS data set designated as "d" and is shown (partially) in Figure 7.8

Figure 7.5. Continuation of the randomization program

```
data c; set b;
    if (1<=pen<=5) then diet='A';
    if (6<=pen<=10) then diet='B';
    if (11<=pen<=15) then diet = 'C';
    if (16<=pen<=20) then diet = 'D';
run;
proc print data=c;
  title1 'Unrandomized Assignment of';
  title2 'Pens to Diets';
run;
proc plan seed=121850;
  factors pen=20;
  output data=c out=d;
  title1 'Random Assignment of 20 Pens';
  title2 'to Four Diets (Balanced)';
proc print data=d;
run;
```

The PLAN procedure can be used to generate research designs involving multiple factors, crossed or nested (see Chapter 13), and also designs involving one or more blocking variables (see Chapter 15). The two-stage randomization procedure illustrated here can be performed more efficiently using the capabilities of the PLAN procedure, once you are more familiar with the issues of designing more complex research studies.

Figure 7.6. A portion of the systematic assignment of pens to diets

```
                Unrandomized Assignment of
                       Pens to Diets

            OBS     CHICK     PEN     DIET

             1        1        1       A
             2        2        7       B
             3        3       13       C
             4        4        1       A
             5        5       19       D
             6        6        4       A
             7        7        1       A
             8        8       17       D
             9        9       19       D
            10       10       16       D
            11       11       18       D
            12       12       11       C
            13       13        6       B
            14       14       20       D
            15       15       18       D
            16       16        8       B
            17       17        9       B
            18       18       10       B
            19       19       16       D
            20       20       10       B
            21       21        2       A
            22       22        5       A
            23       23        7       B
            24       24       16       D
            25       25        8       B
```

Figure 7.7. Random arrangement of pen numbers

```
                Random Assignment of 20 Pens
                  to Four Diets (Balanced)

Procedure PLAN

Factor     Select   Levels    Order
------     ------   ------    -------
PEN           20       20     Random


     PEN
--+--+--+--+--+--+--+--+--+--+--+--+--+--+--+--+--+--+--+--+

 14   7 13 16  9 12  2  4  3 11 20 17  5  6  1  8 19 10 18 15
```

Figure 7.8. A portion of the random assignment of pens to diets

```
                 Random Assignment of 20 Pens
                  to Four Diets (Balanced)

        OBS     CHICK     PEN     DIET

         1        1        14       A
         2        2         2       B
         3        3         5       C
         4        4        14       A
         5        5        18       D
         6        6        16       A
         7        7        14       A
         8        8        19       D
         9        9        18       D
        10       10         8       D
        11       11        10       D
        12       12        20       C
        13       13        12       B
        14       14        15       D
        15       15        10       D
        16       16         4       B
        17       17         3       B
        18       18        11       B
        19       19         8       D
        20       20        11       B
        21       21         7       A
        22       22         9       A
        23       23         2       B
        24       24         8       D
        25       25         4       B
```

7.3 Determining Sample Sizes

Example 7.12. We'll use the hypothesis-testing approach to calculate the sample sizes for the study [to compare the mean frequencies of light emitted by two species of fireflies] described in Example 7.11. Recall that the hypothesis-testing objective of the study was to use a 0.05-level test of $H_0: \theta = 0$ against $H_1: \theta > 0$, where $\theta = \mu_A - \mu_B$ is the difference between the mean light frequencies emitted by two species of fireflies. We want to determine the sample sizes such that, if μ_A is 1 Hz more than μ_B, then the probability of concluding $\mu_A > \mu_B$ is at least 0.80...

It remains to determine a value for σ. Selection of σ is usually based on experience or on some preliminary data collected for this purpose. In the present case, experience with similar data suggests a standard deviation of the measured light frequencies in the range 0.75-0.80 Hz. We will use the conservative value $\sigma = 0.80$.

The iterative procedure illustrated in Rao's Display 7.9 is a job for a computer! SAS does not have any built-in programs for performing this trial-and-error process, but there exists a well-known SAS program, FPOWTAB ("F Power Tables"), written by Ralph G. O'Brien in 1986, which can be used to determine sample size for almost any test of hypothesis involving sample means. A copy of this program is included as Figure 7.13 in the Appendix to this chapter.

Note that on the third line of the FPOWTAB program in Figure 7.13 is the INFILE statement,

```
infile 'powspecs.dat'.
```

In a data file named POWSPECS.DAT ("power specifications") resides the information about the particular hypothesis testing problem at hand. O'Brien's instructions for input to the program are given in the Appendix. Figure 7.9 shows the POWSPECS.DAT file for this two-sample *t*-test.

Lines in Figure 7.9 are referred to as "records". Record 1 contains the title of the analysis and may be up to 78 characters long, counting blank spaces. This exercise is titled, "Power Analysis for Two-Sample *T*-Test."

Figure 7.9. POWSPECS.DAT file for a two-sample *t*-test

```
Power Analysis for Two-Sample T-Test
2 2
1 1 Hz
1 0.05
1 0.80
4 8 9 10 11
Two Means   1 0.50
```

Record 2 contains two entries. The first is the total "basis" sample size, which is the minimum possible size of the two samples combined. In this example, we want the same number of observations for both species A and B, and the smallest total sample size that could satisfy this requirement is for $n_A = 1$ and $n_B = 1$, so that $n_A + n_B = 2$ is the total basis sample size. If it had been deemed necessary to have twice as many observations from species A as from species B, then the smallest total sample size would have been 3, since $n_A = 2$ and $n_B = 1$. The second entry in Record 2 is the number of means in the design, in this case, 2.

Records 3, 4, and 5 allow you to specify several values each for the difference in the population means to be detected, the significance level of the test, and the population standard deviations, respectively. Record 3 first lists the number of different scenarios for differences in the population means that you wish to investigate. Then follow the titles you wish to assign to the scenarios, as many titles as you say you have scenarios to investigate. In Figure 7.9 Record 3 lists that there is only one scenario to be investigated, and that it is being called "1 Hz." If we had wanted to look at the power of the test to detect a 1-Hz. difference *and* also a 2-Hz. difference, then Record 3 would have read

```
2 1 Hz   2 Hz.
```

Note carefully that *two (or more) spaces are used to separate the two titles*, since they are comprised of character values. This program allows up to five different scenarios to be investigated.

Record 4 first lists the number of alpha-levels you want to investigate, and then it lists what they are. The effects of up to five different alpha levels may be compared. In this example, we are only going to consider 0.05-level tests. Had we wanted to compare the sample sizes required for both 0.05-level and 0.01-level tests, Record 4 would have read

```
2 0.05 0.01.
```

In Record 5 are some possible values for the standard deviation of the two populations. Recall that while we are testing to see if the population means differ, we must make the assumption that all populations are *equally variable,* so that the standard deviation applies to all groups being compared. Up to three different values for the standard deviation may be specified: this allows you to examine and compare the effects of "best case," "worst case." and "most likely" amounts of variation on the necessary sample size. In this example, we are considering only one ("worst case") value for the standard deviation, namely 0.80. Had we also wished to see what would happen if the standard deviation were 0.75, then Record 5 would have read

```
                          2 0.75 0.80.
```

In Record 6 are entered first the number and then the values of the multipliers of the basis sample size you wish to consider. For balanced designs (same number of observations in each group), these are simply the number of observations per group. For example, in Figure 7.9, we want to look at the power of the test if we have two groups of 8, 9, 10, and 11 observations each. These can be guessed initially, and then the program can be re-run with refined guesses after examining the power analysis. Alternatively, these multipliers may be arrived at through economic considerations. Up to 5 multipliers can be used.

Subsequent records, unlimited in number, are Effects Records, one for each effect (hypothesis) to be tested. We wish to test solely the difference between the two population means, so we have only one Effects Record, but in studies involving more than two means, for example, there may be several hypotheses of interest. Each Effects Record contains first a title by which the test can be identified on the output. The title may be up to 78 characters long. Then, after at least two spaces is the "numerator degrees of freedom for the effect", which will always be "1" for a two-group comparison. Finally comes a list of the "SSH(Population)" values for each scenario listed in Record 3. This is a function of the difference in the means that is important to detect. For the two-sample *t*-test,

$$SSH(Population) = \Delta^2/2$$

where Δ is the hypothesized or important difference in the means. If in Record 3 we had listed a 2 Hz. scenario in addition to the 1 Hz. scenario, then the Effects Record would read

```
Two Means   1 0.50 2.00.
```

Figure 7.10 shows the output from FPOWTAB. The "Regular F" simply means a two-tailed *t*-test in this instance. Under "1-Tailed", we see that the minimum total sample size required to achieve at least 80% power to detect a difference of 1 Hz. in the population means is 18, so that 9 fireflies of each species need to be observed.

Figure 7.10. Power analysis for the two-sample *t*-test

```
        Power Analysis for Two-Sample T-Test

        Effect: Two Means,
        Degrees of Freedom Hypothesis: 1,
        Scenario: 1 Hz,
        Powers Computed from SSH(Population):
        0.5,
        Using the Basis Total Sample Size: 2,
        AND Total Cells in Design: 2
        ----------------------------------------
        |                        |   Std. Dev   |
        |                        |--------------|
        |                        |     0.8      |
        |                        |--------------|
        |                        |   Total N    |
        |                        |--------------|
        |                        |16 |18 |20 |22 |
        |                        |---+---+---+---|
        |                        |PO-|PO-|PO-|PO-|
        |                        |WER|WER|WER|WER|
        |-----------------------+---+---+---+---|
        |Test Type  |ALPHA       |   |   |   |   |
        |-----------+-----------|   |   |   |   |
        |Regular F  |0.05        |.64|.70|.75|.80|
        |-----------+-----------+---+---+---+---|
        |1-Tailed   |0.05        |.77|.81|.85|.88|
        ----------------------------------------
```

FPOWTAB can also be used to determine the sample size for a one-sample test. Record 2 in this application will always be just

1 1

and in the Effects Records, degrees of freedom will always be "1" and SSH(Population) will be Δ^2, instead of $\Delta^2/2$, where Δ is the difference from the null value which is to be detected.

For example, suppose that Purita Feeds wishes to test a new formulation of their popular Cow Chow. If a mean weight gain of at least 20 lb. more than under the present formulation can be demonstrated, then the new formulation will be deemed worthy of marketing. Past experience with the test herd indicates a standard deviation of weight gains in the neighborhood of 35 lb. How many animals need to be tested if an $\alpha = 0.05$ level test is to have power at least 0.90 of detecting an increase this large?

Figure 7.11 shows the POWSPECS.DAT file containing the specifications for this problem, and the power analysis from the FPOWTAB program is given in Figure 7.12. The indication is that somewhere between 25 and 30 animals need to be used in the experiment. The program could be re-run using multipliers of, say 26, 27, 28, and 29 in Record 6 in order to zero in on the minimum number of animals required.

Figure 7.11. Specification file for a one-sample test

```
Power for One-Sample Test
1 1
1 Twenty Pounds
1 0.05
1 35
5 20 25 30 35 40
One Sample  1 400
```

Figure 7.12. Power analysis for a one-sample test

```
                  Power for One-Sample Test

       Effect: One Sample,
       Degrees of Freedom Hypothesis: 1,
       Scenario: Twenty Pounds,
       Powers Computed from SSH(Population): 400,
       Using the Basis Total Sample Size: 1,
       AND Total Cells in Design: 1
       ------------------------------------------------
       |                          |    Std. Dev     |
       |                          |-----------------|
       |                          |       35        |
       |                          |-----------------|
       |                          |    Total N      |
       |                          |-----------------|
       |                          |20 |25 |30 |35 |40 |
       |                          |---+---+---+---+---|
       |                          |PO-|PO-|PO-|PO-|PO-|
       |                          |WER|WER|WER|WER|WER|
       |--------------------------+---+---+---+---+---|
       |Test Type  |ALPHA         |   |   |   |   |   |
       |-----------+--------------|   |   |   |   |   |
       |Regular F  |0.05          |.68|.78|.86|.91|.94|
       |-----------+--------------+---+---+---+---+---|
       |1-Tailed   |0.05          |.79|.87|.92|.95|.97|
       ------------------------------------------------
```

7.A Appendix: O'Brien's FPOWTAB Program

Figure 7.13 shows the SAS code for performing power analysis for testing hypotheses concerning population means, as presented (slightly modified) in Ralph G. O'Brien (1986): "Power Analysis for Linear Models," *Proceedings of the Eleventh Annual SAS® Users Group International Conference*, pp. 915-922.

Important Notes:

This program is designed so that all specifications (including the title) are entered as data.

There is no need to update the SAS code for each application: one can use an external file ("POWSPECS") of specifications and keep the source file separate (and safe from accidental modification).

Figure 7.13. O'Brien's FPOWTAB program for computing power for testing means

```
options ls=70 ps=54 pageno=1;
data powdata;
infile 'powspecs.dat' eof=last;
file print;
keep scenario alpha power stddev totaln eff_titl df_hypth
     testtype basetotn totcells sshpop;
array alphav(3) alpha1-alpha3;
array scnariov(5) $ 78 scnario1-scnario5;
array sshpopv(5) sshpop1-sshpop5;
array stddevv(3) stddev1-stddev3;
array multfacv(5) multfac1-multfac5;
*********Input Title and All Parameter Values**********;
input titlevar & $78.;
call symput ('mcrtitl',titlevar);
input basetotn totcells;
input num_scn @;
  do i = 1 to num_scn; input scnariov(i) & @; end;
input num_alph @;
  do i = 1 to num_alph; input alphav(i) @; end;
input num_sd @;
  do i = 1 to num_sd; input stddevv(i) @; end;
input num_m @;
  do i = 1 to num_m; input multfacv(i) @; end;
*;
nxteffct:;   *Next Effect ;

input eff_titl & $78. df_hypth @;
  do i = 1 to num_scn; input sshpopv(i) @; end;
*;
*****Loops to Compute All Entries for Table******;
do i_alpha=1 to num_alph; alpha=alphav(i_alpha);
  do i_scn=1 to num_scn; scenario=scnariov(i_scn);
  sshpop=sshpopv(i_scn);
    do i_sd=1 to num_sd; stddev=stddevv(i_sd);
      do i_m=1 to num_m; multfac=multfacv(i_m);
        lambda=multfac*sshpop/stddev**2;
        totaln=multfac*basetotn;
        df_error=totaln-totcells;
        testtype='Regular F';
        fcrit=finv(1-alpha, df_hypth, df_error,0.0);
        power=1-probf(fcrit, df_hypth, df_error, lambda);
```

```
        power=round(power,.01);
        if power gt .99 then power=.99;
        output;
*;
        *Compute power for one-tailed t-tests for single df
              hypotheses;
        if df_hypth=1 then do;
          testtype='1-Tailed T';
          tcrit=tinv(1-alpha, df_error,0);
          power=1-probt(tcrit, df_error, sqrt(lambda));
          power=round(power,.01);
          if power gt .99 then power=.99;
          output;
        end;
*;
    end;
  end;
 end;
end;
go to nxteffct;
*;
last:
*No more effects to input;
run;

proc tabulate format=3.2 order=data;
  class scenario df_hypth eff_titl testtype stddev totaln
      alpha basetotn totcells sshpop;
  var power;
  table eff_titl='Effect:'
    *df_hypth='Degrees of Freedom   Hypothesis:'
    * Scenario = 'Scenario:'
    * sshpop = 'Powers Computed from SSH(Population):'
    * basetotn = 'Using the Basis Total Sample Size:'
    * totcells = 'Total Cells in Design:',
    testtype= 'Test Type' * alpha,
    stddev = 'Std. Dev' *totaln = 'Total N'
      *power*sum=' '/rtspace=25;
  title1 "&mcrtitl";
run;
```

Figure 7.14 shows O'Brien's instructions (slightly modified) for creating the specification file, POWSPECS.DAT.

Figure 7.14. Instructions for entering specifications

*****Instructions for Input to FPOWTAB********

All records are given in list-directed mode: there is no column dependency and records can spill over onto two or more lines. However, each record must begin a new line.

Record 1:   Main title, up to 78 characters
               (2 consecutive spaces mark the end of the title)
  Example:
```
  Input==>  Effect of Diet/Exercise and Drug Therapy on
            LDL Cholesterol
```

Record 2:   Total sample size used as basis for SSH(Population)
               values
               Total number of non-empty cells in design
Example: For a 2×3 factorial with four n=1 cells and two n=2 cells (that is, two cells have twice as many observations as the other four), this record would be:
```
  Input==>  8 6
```

Record 3:   Number of scenarios for population means (up to 5)
               Title of first scenario (up to 78 characters)
               Title of second scenario, if any
               Etc.
               Title of last scenario
               (2 or more consecutive spaces mark the ends of titles)
  Example:
```
  Input==>  2 Low Effect  High Effect
```

Record 4:   Number of alpha levels (up to 3)
               First value for alpha
               Second value, if any
               Third value, if any
  Example:

```
   Input==>   2 0.01 0.05
```

Record 5:   Number of standard deviations (up to 3)
               First value of standard deviation
               Second value, if any
               Third value, if any

  Example:
```
   Input==>   2 0.1 0.2
```

Record 6:   Number of multiplicative factors for total *n* (up to 5)
               First value for multiplicative factor
               Second value for multiplicative factor, if any
               Etc.
               Last value for multiplicative factor

  Example: The basis total sample size is 8. Therefore, in order to get
power for total sample sizes of 104, 208, and 416, use:
```
   Input==>   3 13 26 52
```

Effects Records.   One record for each hypothesis (effect) considered,
        unlimited number. Each record has the following format:
               Title, up to 78 characters, followed by at least t 2 spaces
               Degrees of freedom (numerator) for effect
               SSH(Population) value for first scenario of means
               SSH(Population) value for second scenario, if any
               Etc.
               SSH(Population) value for last scenario

  Example: Six effects and two scenarios.   SSH(Pop) values are
computed based on specified values of means or specified differences
between means. (See text.)
```
   Input==>   Diet/Exercise Main Effect  1 0.002 0.00288
              Drug Main Effect  2 0.00674 0.01504
              Diet/Ex by Drug Interact  2 0.00034 0.00104
              Drug(2 -1 -1)  1 10.00551 0.01201
              Drug(0 1 -1)  1 0.00123 0.00303
              Drug(2 -1 -1) in Diet(2)  1 0.0016 0.00303
```

8

Single-Factor Studies: One-Way ANOVA

8.1 Introduction

Either the GLM (general linear model) procedure or the ANOVA (analysis of variance) procedure may be used to analyze a single-factor experiment. The ANOVA procedure is a subset of the methods available in the GLM procedure, and is more efficient than GLM for either single-factor analysis of variance data or *balanced* multifactor analysis of variance data. (to be addressed in Chapter 13). While many of the same commands can be used in both ANOVA and GLM, there are some instances in which commands are available only in GLM.

The PLOT and UNIVARIATE procedures are used in conjunction with ANOVA and GLM to give a more complete insight into the data.

The NPAR1WAY procedure is used to compare group averages when the responses are ordinal, and the FREQuency procedure can be used to compare groups when the response is categorical.

8.2 Completely Randomized Designs

Example 8.2. Photomorphogenetic studies of plants are often conducted under a light condition called safelight, whose effect on certain plant properties is the same as the effect of darkness. The effect of two sources of safelight (A and B), each at two intensities (low and high), was compared with the effect of darkness (D) in the following experiment.

The experimenter chose 20 identical seedlings. Four randomly selected seedlings were grown under each of $t = 5$ treatments: D, AL (source A at low intensity), AH (source A at high intensity), BL (source B at low intensity) and BH (source B at high intensity). The heights (cm) of the twenty plants were measured after four weeks.

Figure 8.1 shows the SAS program to analyze the data of Example 8.2 using the ANOVA procedure.

Figure 8.1. SAS program using the ANOVA procedure for a single-factor study

```
options ls=70 ps=54;
/*Example 8.2.   Single-factor ANOVA*/
data;
  input tmt $ @;
    do i = 1 to 4;
      input height @;
       output;
    end;
  drop i;
lines;
 D 32.94 35.98 34.76 32.40
AL 30.55 32.64 32.37 32.04
AH 31.23 31.09 30.62 30.42
BL 34.41 34.88 34.07 33.87
BH 35.61 35.00 33.65 32.91
;
run;
proc anova;
   class tmt;
   model height = tmt;
   title 'Using ANOVA to Analyze a Single-
       Factor Study';
run;
```

The only change in the program in Figure 8.1 necessary to use the GLM procedure instead of the ANOVA procedure to analyze these data is to simply write

```
proc glm;
```

instead of

```
proc anova;.
```

(Presumably, you would also want to change the title, although this will not affect the analysis.) We will see that the output looks somewhat different for the two procedures, however.

Note the use of the CLASS statement in the ANOVA (or the GLM) procedure. As was the case in the TTEST procedure, this statement designates to SAS that the CLASS variable is a categorical variable that defines the groups whose means we want to compare. In this example, the four types of safelight were entered in the data set as character values (D, AL, etc.); however, one might choose to enter the treatment number (1, 2, etc.) instead. The CLASS statement designates that even if the groups are designated by numbers, the variable defining the groups is still a *categorical* variable.

Figure 8.2. Class Levels Information

```
           Using ANOVA to Analyze a Single-Factor Study

                   Analysis of Variance Procedure
                      Class Level Information

              Class     Levels    Values

              TMT          5       AH AL BH BL D


         Number of observations in data set = 20
```

Use of the CLASS statement in ANOVA or GLM produces a page of output titled "Class Levels Information," as shown in Figure 8.2. This is a handy place to check that SAS has read your input statement correctly and has read all 20 data values and the five different factor levels (and that you didn't make a typographical error in entering the factor levels). Note carefully that the factor levels are listed in Figure 8.2 in *alphabetical* order, even though they were not entered that way in the program. We will see later, when we get to comparing the mean responses at each of the treatments with each other, that this is a very important fact for you to know. Had the factor levels been inputted as treatment numbers instead of as

character values, they would appear in *numerical* order in the Class Levels
Information.

GLM and ANOVA both require a MODEL statement of the form

```
model dep = indep;
```

where *dep*, the numerical dependent variable is listed on the left-hand side
and the categorical independent, or class, variable appears to the right of the
equals sign.

Figure 8.3. Analysis of variance from the ANOVA procedure

```
               Using ANOVA to Analyze a Single-Factor Study

                     Analysis of Variance Procedure

Dependent Variable: HEIGHT

Source                  DF      Sum of Squares    F Value      Pr > F

Model                    4         41.08077000      9.41       0.0005

Error                   15         16.37655000

Corrected Total         19         57.45732000

                  R-Square              C.V.            HEIGHT Mean

                  0.714979            3.159404            33.0720000


Source                  DF            Anova SS    F Value      Pr > F

TMT                      4         41.08077000      9.41       0.0005
```

Figure 8.3 shows the analysis of variance from the program in
Figure 8.1. The ANOVA table in Rao's Example 8.3 can be constructed
using the information in Figure 8.3. The total amount of variation in heights
is measured by the (corrected) total sum of squares,

$$SS[TOT] = 57.45732$$

with 19 degrees of freedom. This total amount of variation is broken down
into variation explained by the model

$$SS[TOT] = 41.08077$$

with 4 degrees of freedom (number of treatments minus 1) and unexplained variation,

$$SS[E] = 16.37655$$

with 15 d.f. Thus, differences in the five treatment conditions explains

$$R^2 = SS[T]/SS[TOT]$$

$$= 0.714779,$$

or about 71.5% of the variation in the plant heights.

Dividing the sums of squares by their corresponding degrees of freedom produces mean squares. Mean squares are not shown in Figure 8.3 due to space limitations implied by the LINESIZE=70 option. Had the linesize been 72 or more, the analysis of variance table would have shown

$$MS[T] = 41.08077/4 = 10.27019$$

and

$$MS[E] = 16.37655/15 = 1.09177.$$

The square root of the error mean square is given in Figure 8.3 as

```
Root MSE

1.044878
```

which is an estimate of the plant-to-plant within-treatment variability in height (σ^2 in the notation of Rao's text). The mean plant height for all 20 plants is

```
HEIGHT Mean

33.0720
```

and the ratio of the Root MSE to the HEIGHT Mean (multiplied by 100) is the coefficient of variation

```
C. V.

3.159404
```

which indicates that the standard deviation of the plant heights is about 3% of the mean height.

The *F*-value in Figure 8.3 is the ratio of MS[T] to MS[E]:

$$F = 10.27109/1.09177$$

$$= 9.41.$$

This indicates that the treatment-to-treatment variation in this experiment is more than nine times greater than the plant-to-plant variation. The probability of this being just a chance occurrence is quite small,

$$p = 0.0005,$$

giving quite strong evidence of differences in mean heights attributable to differences in the experimental treatments. In Chapter 9 we will look at how to use SAS to investigate which treatments differ from others.

Using the GLM procedure to analyze these data will produce class levels information identical to that in Figure 8.2. The analysis of variance from GLM looks like Figure 8.4. The top part of the output looks identical to the ANOVA output, but the GLM output also includes Type I and Type III analyses. The Type I and Type III analyses produced by GLM are helpful when analyzing unbalanced multifactor studies, as we will see in Chapter 13.

Figure 8.4. Analysis of variance from GLM procedure

```
                 Using GLM to Analyze a Single-Factor Study

                    General Linear Models Procedure

Dependent Variable: HEIGHT

Source                  DF    Sum of Squares    F Value     Pr > F

Model                    4       41.08077000      9.41      0.0005

Error                   15       16.37655000

Corrected Total         19       57.45732000

              R-Square              C.V.              HEIGHT Mean

              0.714979            3.159404            33.0720000


Source                  DF       Type I SS     F Value     Pr > F

TMT                      4       41.08077000      9.41      0.0005

Source                  DF      Type III SS    F Value     Pr > F

TMT                      4       41.08077000      9.41      0.0005
```

8.3 Checking for Violations of Assumptions

Detecting Violations

In order to check that the residuals from an analysis of variance are normally distributed, we can use the UNIVARIATE procedure to obtain a normal q-q plot, often called a normal probability plot. A plot of residuals versus predicted values (treatment means, in this case) is helpful in visually checking for equality of variances from one treatment to another. To obtain these analyses, the predicted and residual values from the analysis of variance must be outputted to a SAS data set. Figure 8.5 shows in boldface the modifications necessary to the GLM (or ANOVA) program. Predicted values are called "ybar", residuals are named "resid", and they are outputted to a data set called "check".

Figure 8.5. GLM program to check validity of ANOVA assumptions

```
options ls=70 ps=54;
/*Example 8.2.  Single-factor ANOVA*/
data;
  input tmt $ @;
    do i = 1 to 4;
      input height @;
       output;
    end;
  drop i;
lines;
 D 32.94 35.98 34.76 32.40
AL 30.55 32.64 32.37 32.04
AH 31.23 31.09 30.62 30.42
BL 34.41 34.88 34.07 33.87
BH 35.61 35.00 33.65 32.91
;
run;
proc glm;
  class tmt;
  model height = tmt;
  output out=check p=ybar r=resid;
  title 'Using GLM to Analyze a Single-Factor
      Study';
run;
proc plot data=check;
  plot resid*ybar/vref=0;
run;
proc univariate plot data=check;
  var resid;
run;
```

The PLOT procedure produces a scatter diagram. In the PLOT command, the first variable listed will be plotted on the vertical axis, and the one following the asterisk becomes the horizontal axis variable. Single points are plotted with the letter "A", two coincident points are denoted by the letter "B", three coincident points with a "C", and so on. (Keep in mind that this is a rather crude plot, being drawn essentially with a typewriter. It is possible to produce very nice graphs using SAS, but often these crude ones are sufficient for personal work, if not suitable for publication.) The option

Figure 8.6. Plot of residuals against treatment means

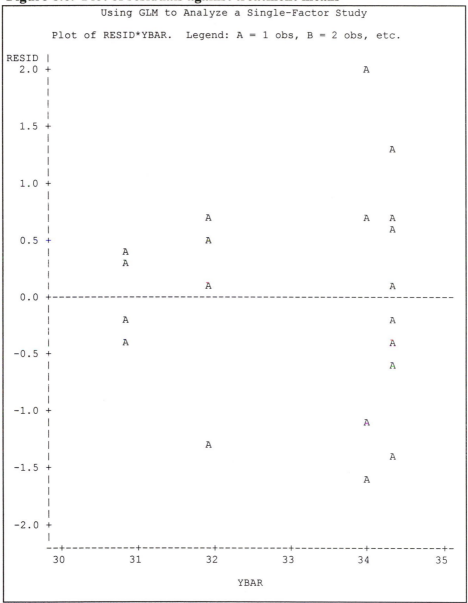

Figure 8.7. Normality check from UNIVARIATE

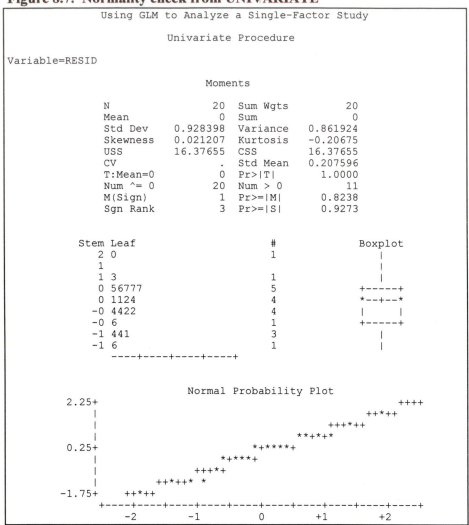

```
                 Using GLM to Analyze a Single-Factor Study

                          Univariate Procedure

Variable=RESID

                                Moments

              N                20  Sum Wgts            20
              Mean              0  Sum                  0
              Std Dev    0.928398  Variance      0.861924
              Skewness   0.021207  Kurtosis      -0.20675
              USS        16.37655  CSS           16.37655
              CV                .  Std Mean      0.207596
              T:Mean=0          0  Pr>|T|          1.0000
              Num ^= 0         20  Num > 0             11
              M(Sign)           1  Pr>=|M|         0.8238
              Sgn Rank          3  Pr>=|S|         0.9273


       Stem Leaf                        #           Boxplot
          2 0                           1              |
          1                                            |
          1 3                           1              |
          0 56777                       5           +-----+
          0 1124                        4           *--+--*
         -0 4422                        4           |     |
         -0 6                           1           +-----+
         -1 441                         3              |
         -1 6                           1              |
            ----+----+----+----+


                        Normal Probability Plot
       2.25+                                            ++++
           |                                        ++*++
           |                                    +++*++
           |                                 **+*+*
       0.25+                        *+*****+
           |                    *+****+
           |                +++*+
           |          ++*++* *
      -1.75+    ++*++
           +----+----+----+----+----+----+----+----+----+----+
               -2        -1         0        +1        +2
```

vref=0

after the slash in the PLOT command asks that a reference line be drawn at
the value zero on the vertical axis.

Figure 8.6 shows the plot of residuals against treatment means. It appears as if the variance might be proportional to the mean, so that a log transformation might be appropriate. (See Rao, Section 8.7.) Relevant portions of the output from UNIVARIATE are shown in Figure 8.7. Both the stem-and-leaf diagram and the normal probability plot appear to be reasonably "well-behaved." indicating no striking departure from normality.

Finally, adding

```
means tmt;
```

to the list of GLM or ANOVA commands will produce the table of treatment means and standard deviations shown in Figure 8.8. The treatment standard deviations can be used to perform the Bartlett test for equality of variances.

Figure 8.8. Treatment means and variances

```
            Using GLM to Analyze a Single-Factor Study

               General Linear Models Procedure

        Level of          ------------HEIGHT-----------
        TMT        N       Mean              SD

        AH         4      30.8400000        0.38270964
        AL         4      31.9000000        0.93284511
        BH         4      34.2925000        1.23294701
        BL         4      34.3075000        0.44199359
        D          4      34.0200000        1.65126214
```

Transformations

Example 8.9. In a study to compare the effects of three diets on the quality of chicken eggs, 24 birds were randomly divided into three groups of eight birds and assigned to one of three diets: diet 1, diet 2 and diet 3. The numbers of eggs (out of a total of 25) classified as grade A for each bird are as follows…

These data can be regarded as the numbers of successes in binomial experiments with $n = 25$ trials (grading of the 25 eggs laid by each chicken); the probability of success is equal to the probability of laying a grade A egg...

Clearly, a one-way ANOVA of the raw data is inappropriate, because the populations are not normal (the samples are from binomial populations) and the population variances may not be equal. However, if we apply the *arcsine* transformation to the data, then a one-way ANOVA can be used to compare the means of the transformed populations.

In his Section 8.7, Rao discusses the use of the logarithmic, arcsin, square root, and power transformations. The SAS functions to apply these transformations are given in Figure 8.9. These functions are used in a data step, as in, for example,

$$y = sqrt(x);$$

where the variable X was defined in an input statement.

Figure 8.9. SAS functions for common transformations

Transformation	SAS Function
$\log_{10}$	log10($\cdot$)
ln	log($\cdot$)
arcsine	arsin($\cdot$)
square root	sqrt($\cdot$)
$(\cdot)^{\lambda}$	$(\cdot)**\lambda$

The program in Figure 8.10 illustrates the use of the arcsine transformation in testing the null hypothesis that the three diets in Example 8.9 all yield the same proportion of grade A eggs.

The printout of the first few cases of the transformed data, shown in Figure 8.11, does not give the same transformed values as in Rao's Example 8.9. This is because Rao's values are expressed in degrees, while SAS uses radians. This is no problem, however, because the F-test is unaffected by whether data are expressed in degrees or radians, as shown by comparing the ANOVA table in Figure 8.12, with Rao's. (To convert to degrees, if desired, multiply radians by 57.2957.)

Figure 8.10. Using the arcsine transformation

```
options ls=70 ps=54;
data diets;
   input diet @@;
   do I = 1 to 8;
      input aeggs @@;
      output;
   end;
lines;
1 21 18 21 24 22 20 21 20
2 20 17 19 14 19 12 16 11
3 22 24 21 22 24 23 21 23
;
run;
data b; set diets;
fraction = aeggs/25;
rootf = sqrt(fraction);
teggs = arsin(rootf);
run;
proc print data =b;
run;
proc anova data = b;
   class diet;
   model teggs = diet;
run;
```

Figure 8.11. Partial listing of transformed values

OBS	DIET	I	AEGGS	FRACTION	ROOTF	TEGGS
1	1	1	21	0.84	0.91652	1.15928
2	1	2	18	0.72	0.84853	1.01320
3	1	3	21	0.84	0.91652	1.15928
4	1	4	24	0.96	0.97980	1.36944
5	1	5	22	0.88	0.93808	1.21705
6	1	6	20	0.80	0.89443	1.10715
7	1	7	21	0.84	0.91652	1.15928
8	1	8	20	0.80	0.89443	1.10715

Figure 8.12. **ANOVA table illustrating use of the arcsine trans-formation.**

Analysis of Variance Procedure				
Dependent Variable: TEGGS				
Source	DF	Sum of Squares	F Value	Pr > F
Model	2	0.44677287	17.67	0.0001
Error	21	0.26552945		
Corrected Total	23	0.71230232		

8.4 One-Way Classifications with Ordinal Data

Example 8.11. To compare the resistance of a group of bacteria to an antibiotic, investigators tested five specimens of each of three strains against concentrations of the antibiotic and determined the lowest concentration that inhibited visible bacteria growth (the minimal inhibitory concentration).

The Kruskal-Wallis test is an extension of the Wilcoxon rank sum test to more than two groups. The Kruskal-Wallis test can be obtained using exactly the same code as for the Wilcoxon test: if there are two groups, the NPAR1WAY procedure will give the Wilcoxon test; if there are more than two groups, the procedure will give the Kruskal-Wallis test. Figure 8.13 shows the SAS program to analyze these data.

Figure 8.14 shows the Kruskal-Wallis test as applied to these data. As was apparent from a visual inspection of the data, we have strong evidence ($p = 0.0053$) that the three strains differ. Note that SAS uses a Chi-Square approximation to the Kruskal-Wallis test. This test works well in large samples, in that it gives a good approximation to the accurate p-value.

Figure 8.13. Using the NPAR1WAY procedure to perform the Kruskal-Wallis test

```
options ls=70 ps=54;
/*Example 8.11.  Kruskal-Wallis test*/
data;
input strain @;
  do i = 1 to 5;
    input concen @;
    output;
  end;
drop i;
lines;
1 20 25 30 28 26
2  8 10  2  8 11
3 40 35 30 20 30
;
run;
proc npar1way wilcoxon;
  class strain;
  var concen;
run;
```

Figure 8.14. Kruskal-Wallis test

```
              N P A R 1 W A Y   P R O C E D U R E

      Wilcoxon Scores (Rank Sums) for Variable CONCEN
                Classified by Variable STRAIN


                        Sum of    Expected      Std Dev        Mean
    STRAIN      N        Scores    Under H0      Under H0      Score

    1           5   45.5000000        40.0     8.12110713   9.1000000
    2           5   15.0000000        40.0     8.12110713   3.0000000
    3           5   59.5000000        40.0     8.12110713  11.9000000

               Average Scores were used for Ties
        Kruskal-Wallis Test (Chi-Square Approximation)
         CHISQ=  10.467    DF=  2    Prob > CHISQ=      0.0053
```

8.5 One-Way Classification with Nominal Data

Example 8.14. To determine whether the distribution of blood types is different in three ethnic subgroups of a population, blood samples from 100 subjects within each ethnic group were classified according to blood type.

In Chapter 6, two populations of categorical responses were compared in terms of their distributions by means of a Chi-Square test. The same test extends easily to comparing more than two distributions, as shown in the program in Figure 8.15 and the output in Figure 8.16. The Chi-Square

Figure 8.15.   The Chi-Square test to compare three categorical distributions

```
options ls=70 ps=54;
/*Example 8.14.  One-Way classification:
Categorical data */
data;
  input group $ @;
    do i = 1 to 4;
      input type count @;
      output;
    end;
 drop i;
lines
;
   I 1 44 2 31 3 14 4 11
  II 1 38 2 39 3 10 4 13
III 1 56 2 26 3 12 4  6
;
run;
proc format;
  value type 1 = 'O' 2 = 'A' 3 = 'B' 4 =
      'AB';
run;
proc freq;
  tables group * type/ chisq nocol norow
      nopercent expected;
  weight count;
  format type type.;
run;
```

value in Figure 8.16 is significant only at $p = 0.142$, so we do not have convincing evidence that the three subpopulations differ in their blood type distributions.

In addition to the Chi-Square test, SAS provides two other tests as standard output from the FREQ procedure. The Likelihood Ratio Chi-Square is a more general form of the Chi-Square test and will usually give a value of the test statistic and *p*-value similar to the Chi-Square. The Mantel-Haenszel statistic is used to test for a linear association (see Chapter 10) between the row and column variables and is meaningless unless both of the variables are at least ordinal.

Figure 8.16. Comparing three categorical distributions

```
                         TABLE OF GROUP BY TYPE

         GROUP     TYPE

         Frequency|
         Expected |O       |A       |B       |AB      |   Total
         ---------+--------+--------+--------+--------+
         I        |    44  |    31  |    14  |    11  |    100
                  |    46  |    32  |    12  |    10  |
         ---------+--------+--------+--------+--------+
         II       |    38  |    39  |    10  |    13  |    100
                  |    46  |    32  |    12  |    10  |
         ---------+--------+--------+--------+--------+
         III      |    56  |    26  |    12  |     6  |    100
                  |    46  |    32  |    12  |    10  |
         ---------+--------+--------+--------+--------+
         Total        138       96       36       30       300


                STATISTICS FOR TABLE OF GROUP BY TYPE

         Statistic                   DF      Value      Prob
         ------------------------------------------------------
         Chi-Square                   6      9.606      0.142
         Likelihood Ratio Chi-Square  6      9.723      0.137
         Mantel-Haenszel Chi-Square   1      2.989      0.084
         Phi Coefficient                     0.179
         Contingency Coefficient             0.176
         Cramer's V                          0.127

         Sample Size = 300
```

9

Comparison of Population Means and Determination of Sample Sizes in Single Factor Studies

9.1 Introduction

Once the F-test in the ANOVA or GLM procedure has detected that the treatment means differ, the next step is to determine which means differ from which others. In this chapter, we learn how to get SAS to test or estimate contrasts among the treatment means and also look at some of the procedures available for making all pairwise comparisons among treatment means.

In addition, a technique for reconstructing a data set from just the summary statistics is shown.

9.2 Making Specified *A Priori* Comparisons

Example 9.1 The reduction in systolic blood pressure (BP) after a drug for hypertension is administered is one of the key indicators of how well the patient is responding to the drug. When treating for hypertension the side effects associated with the drug are of particular concern. In a study two drugs, A and B, for reducing the side effects of a standard hypertension drug S were evaluated. Drugs A and B were administered concurrently with drug S. The study was conducted using a completely randomized design with five treatments as follows:

Treatment	Drug
1	Standard drug (S)
2	S combined with a low dose of A (S + AL)
3	S combined with a high dose of A (S + AH)
4	S combined with a low dose of B (S + BL)
5	S combined with a high dose of B (S + BH)

There were four replications of each treatment. The reduction in blood pressure (mm Hg) over a period of four weeks observed for the experimental subjects may be tabulated as follows....

The following are three of the questions that the investigators asked about the treatment effects.

1. Is there a difference between effects of low and high doses of A? In other words, on average, is the reduction in systolic BP the same for S + AL the same and for S + AH? In practical terms, are the expected responses for treatment 2 and treatment 3 different?

2. Is there a difference between the effects of high and low doses of B? In other words, are the expected responses for treatment 4 and treatment 5 different?

3. Is there a difference between the average of the expected responses for the two doses of A (S + AL and S + AH) and the average of the expected responses for the two doses of B (S + BL and S + BH)? (See also Example 9.4.)

Example 9.2. Let's first use the information in Box 9.2 to construct a 95% confidence interval for the expected drop in blood pressure (BP) for a patient treated with the standard drug (treatment 1).

Example 9.3. A principal objective of the BP reduction study was to see how the effects of four combination therapies--namely S + AL, S + AH, S + BL and S + BH--compare with the effect of the standard therapy. In this example, let's use the data to see if there is reason to believe that the expected BP reduction for S + AH is significantly lower than that for the standard therapy.

Example 9.5. Let $\bar{Y}_1, \ldots, \bar{Y}_5$ denote the observed mean responses for the five treatments in Example 9.1. Consider the linear combinations

$$\hat{\theta}_1 = \bar{Y}_1 - \frac{1}{4}(\bar{Y}_2 + \bar{Y}_3 + \bar{Y}_4 + \bar{Y}_5)$$

$$\hat{\theta}_2 = \bar{Y}_2 - \bar{Y}_3$$

$$\hat{\theta}_3 = \bar{Y}_4 - \bar{Y}_5$$

$$\hat{\theta}_4 = \frac{1}{2}(\bar{Y}_2 + \bar{Y}_3) - (\bar{Y}_4 - \bar{Y}_5)$$

Is $\{\hat{\theta}_1, \hat{\theta}_2, \hat{\theta}_3, \hat{\theta}_4\}$ a mutually orthogonal set of comparisons?

Example 9.6. Let's determine the significance of the comparison $\hat{\theta}_1$ in Example 9.5 and interpret the result.

Example 9.7. In Example 9.2 we presented the ANOVA table for a one-way analysis of BP reduction data. In this example let's decompose the treatment sum of squares using the set of orthogonal contrasts described in Example 9.5.

If, as in Example 9.2, the objective is simply to estimate the mean response under one of the treatment conditions, either the ANOVA or GLM procedures with a MEANS statement may be used to obtain an analysis of variance table and a table of treatment means. Figure 9.1 displays the SAS program using the GLM procedure to obtain an analysis of variance and treatment means for the data of Example 9.1. These results are shown in Figures 9.2 and 9.3, respectively.

Note that instead of using numerical designations for the five conditions, as in Rao, the variable TMT is entered as a character variable. This results in the means in Figure 9.3 being listed in *alphabetical* order. Had the conditions been designated numerically, then the means would have been listed in numerical order.

Figure 9.1. Program for analyzing BP data

```
options ls=70 ps=54;
data bp;
  input tmt$ @@;
  do i = 1 to 4;
    input reductn @@;
    output;
  end;
lines;
S    27 26 21 26
AL 19 13 15 16
AH 15 10 10 11
BL 22 15 21 18
BH 20 18 17 16
;
run;
proc glm data = bp;
  class tmt;
  model reductn = tmt;
  means tmt;
run;
```

Figure 9.2. ANOVA table for BP data

```
                General Linear Models Procedure

Dependent Variable: REDUCTN

Source                DF    Sum of Squares    F Value    Pr > F

Model                  4      388.70000000     15.10     0.0001

Error                 15       96.50000000

Corrected Total       19      485.20000000

              R-Square             C.V.           REDUCTN Mean

              0.801113          14.24945             17.8000000
```

From the analysis of variance in Figure 9.2,

$$MS[E] = SS[E]/df_E$$
$$= 96.5/15$$
$$= 6.4333$$

Figure 9.3. Treatment means and standard deviations for BP data

```
              General Linear Models Procedure

        Level of        -----------REDUCTN-----------
        TMT        N       Mean              SD

        AH         4     11.5000000        2.38047614
        AL         4     15.7500000        2.50000000
        BH         4     17.7500000        1.70782513
        BL         4     19.0000000        3.16227766
        S          4     25.0000000        2.70801280
```

so a 95% confidence interval on μ_S is given by

$$25.00 \pm 2.131 \sqrt{\frac{6.4333}{4}}$$

$$= (22.297, 27.703)$$

since $t(15, 0.025) = 2.131$.

      In order to write contrasts among the treatment means, refer to the Class Levels Information portion of the output from the GLM or ANOVA procedure. Recall that if the levels of the factor are entered as numbers, they will be listed in numerical order, but if factor levels are given as character values, they will be listed in alphabetical order, *regardless of the order in which they occur in the data set*. Figure 9.4, which gives the Class Levels Information for the BP example, shows the five treatments listed in alphabetical order. Therefore, even though in Figure 9.1 the responses under the standard treatment (S) were the first ones entered in the data set, SAS is calling Treatment AH = Treatment 1, Treatment AL = Treatment 2, and so on, with Treatment S = Treatment 5. If you want the standard drug alone to be Treatment 1, then you will have to input its value as the number 1 instead of the letter S, and similarly for the other treatments. Keep this in mind when comparing these results to Rao's.

Figure 9.4. Class levels information for BP data

```
                  Class Level Information

         Class      Levels    Values

         TMT            5      AH AL BH BL S


        Number of observations in data set = 20
```

In order to test or estimate contrasts among treatment means, data must be analyzed using the *GLM--not ANOVA--*procedure. If the objective is to simply test the significance of a contrast among means, this can be accomplished by including a

```
                        contrast
```

statement in the GLM procedure. In order to estimate the magnitude of the difference between the groups of means in the contrast, the

```
                        estimate
```

command is used. The former produces a sum of squares for the contrast with its associated degrees of freedom, and *F*-statistic and *p*-value for testing the null hypothesis that the contrast is zero. The latter produces a numerical estimate of the contrast along with its standard error, and the value of the *t*-statistic with associated *p*-value for testing the null hypothesis that the contrast is zero.

Whether you choose to use the ESTIMATE or the CONTRAST command, the form is the same. Following the command, provide in single quotes a short descriptive title of what the contrast is about. This title is *not taken as instructions* by SAS, but will simply provide a label by means of which you can identify the output. Follow this by the name of the factor whose levels you want to compare. Then list the contrast coefficients on the means, *in the order the corresponding groups are listed in the Class Levels Information table.*

For example, to compare the mean response under condition AL with that under AH, first write either

```
contrast or estimate
```

and then, enclosed in single quotes, a descriptive title such as

```
'ALvsAH'
```

followed by the name of the factor whose levels are to be compared

```
tmt
```

in this case. Then list, *in the order AH, AL, BH, BL, and S,* the contrast coefficients, −1, 1, 0, 0, and 0, respectively, to estimate the contrast

$$\mu_{AL} - \mu_{AH} \, .$$

The statement becomes

```
contrast 'ALvsAH' tmt  -1 1 0 0 0;
```

Suppose that instead of entering treatments as a character variable, we had designated the five conditions by the numerical indexes, 1, 2, 3, 4 and 5, given in Example 9.1. Then in order to make the contrast

$$\mu_{AL} - \mu_{AH} \, ,$$

the command would have been

```
contrast '2vs3' tmt 0 1 -1 0 0;
```

since now the coefficients correspond to the numerical, rather than the alphabetical, ordering.

The partial program in Figure 9.5 shows how to use both the CONTRAST and the ESTIMATE commands in the GLM procedure to test

Figure 9.5. GLM program for performing contrasts and estimates

```
proc glm data = bp;
   class tmt;
   model reductn = tmt;
   means tmt;
   estimate 'ALvsAH' tmt   -1     1     0     0
       0;
   estimate 'BLvsBH' tmt    0     0   - 1     1
       0;
   estimate 'AvsB'   tmt  -.5   -.5    .5    .5
       0;
   estimate 'SvsAH'  tmt   -1     0     0     0
       1;
   estimate 'SvsOth' tmt -.25  -.25  -.25  -.25
       1;
   contrast 'ALvsAH' tmt   -1     1     0     0
       0;
   contrast 'BLvsBH' tmt    0     0   - 1     1
       0;
   contrast 'AvsB'   tmt  -.5   -.5    .5    .5
       0;
   contrast 'SvsAH'  tmt   -1     0     0     0
       1;
   contrast 'SvsOth' tmt -.25  -.25  -.25  -.25
       1;
run;
```

Figure 9.6. Contrasts and estimates for BP example

```
                    General Linear Models Procedure

Dependent Variable: REDUCTN

Contrast              DF        Contrast SS       F Value      Pr > F

ALvsAH                 1        36.12500000          5.62      0.0316
BLvsBH                 1         3.12500000          0.49      0.4965
AvsB                   1        90.25000000         14.03      0.0019
SvsAH                  1       364.50000000         56.66      0.0001
SvsOth                 1       259.20000000         40.29      0.0001

                               T for H0:    Pr > |T|    Std Error of
Parameter         Estimate    Parameter=0               Estimate

ALvsAH           4.2500000          2.37      0.0316     1.79350681
BLvsBH           1.2500000          0.70      0.4965     1.79350681
AvsB             4.7500000          3.75      0.0019     1.26820083
SvsAH           13.5000000          7.53      0.0001     1.79350681
SvsOth           9.0000000          6.35      0.0001     1.41789163
```

and estimate the contrasts mentioned in Examples 9.1, 9.3 and 9.6. The contrasts and estimates resulting from this program are shown in Figure 9.6.

Let us consider first the lower portion of Figure 9.6, in which the estimates of the contrasts are given. As in Example 9.3, the comparison between the mean reduction in BP under the standard drug with that for condition AH shows that the difference is estimated to be 13.5, with a standard error of 1.7935. The t-statistic shows that 13.5 is estimated to be 7.53 standard deviations from zero. The probability of seeing a difference this large purely by chance is only (less than) $p = 0.0001$, so we have quite strong evidence that the mean reduction in BP under the standard drug differs from that under the standard drug combined with a high level of drug A. If the research hypothesis is specifically that the mean under the AH is *less* than that under the standard drug, then

$$p = 0.0001/2$$

$$= 0.00005,$$

and we have strong evidence that the mean under AH is less than that under S. Note that the contrast was written in Figure 9.5 so as to form the difference

$$\mu_S - \mu_{AH}$$

so that μ_S has coefficient $+1$, μ_{AH} has coefficient -1, and all other treatment means are assigned coefficients zero in the statement

```
estimate 'SvsAH' tmt -1 0 0 0 1;
```

Had the command been written as

```
estimate 'SvsAH' tmt 1 0 0 0 -1;,
```

then the difference

$$\mu_{AH} - \mu_S$$

would have been estimated to be −13.5 and the *t*-value given as −7.53.

Like information is given about the other four contrasts. For any of the contrasts, confidence interval estimates can be obtained by giving each estimate a margin for error. The margin for error is obtained by multiplying the standard error for the contrast by a *t*-value appropriate for the confidence level with degrees of freedom equal to degrees of freedom for error in the ANOVA table.

Note that in Figure 9.5, when writing either ESTIMATE or CONTRAST commands, it is permissible to leave off trailing zero coefficients. That is, the comparison between high and low doses of drug A could have been written as

```
estimate 'ALvsAH' tmt -1 1;
```

instead of as

```
estimate 'ALvsAH' tmt -1 1 0 0 0;.
```

However, leaving off the *initial* zeroes is not permissible. For instance, the statement

```
estimate 'BLvsBH' tmt -1 1 0;
```

obtained by leaving off the leading zeroes in the command for comparing BL to BH actually instructs SAS to compare *AL and AH*, regardless of the title assigned to the contrast.

The contrast comparing results under the standard treatment to the average under the other four conditions was

```
estimate 'SvsOth' tmt -.25 -.25 -.25 -.25 1;
```

It could also have been written as

```
estimate 'SvsOth' tmt -1 -1 -1 -1 4;
```

and this would be a valid contrast which makes the same comparison as the one in Figure 9.5. However, since the coefficients in this latter version are four times the coefficients in the original version, the estimate and its standard error will also be four times as large. Should you want to use these integer coefficients but estimate the difference between the *average* of the four experimental conditions and the standard one, write

```
estimate 'SvsOth' tmt -1 -1 -1 -1 4/divisor=4; .
```

The use of this DIVISOR option in the ESTIMATE command is very handy when rounding error prevents the coefficients in a contrast from summing to zero. For example, to compare the average heights under BH and BL to the average under the other three treatments, we would like to estimate

$$\frac{\mu_{BL} + \mu_{BH}}{2} - \frac{\mu_S + \mu_{AH} + \mu_{AL}}{3}.$$

To do this, consider writing the statement

```
estimate 'BvsOth' tmt -.33 -.33 .5 .5 -.33; .
```

Unfortunately, SAS will not recognize this as a contrast, because the coefficients do not sum to zero, exactly, no matter how many decimal places you use, and will give you an error message in the LOG file. However, using the DIVISOR option, the desired contrast can be estimated by

```
estimate 'BvsOth' tmt -2 -2 3 3 -2/divisor=6;
```
or
```
estimate 'BvsOth' tmt -1 -1 1.5 1.5 -1/divisor=3;
```

or something else algebraically equivalent.

The information given by the CONTRAST commands is complementary to that given by ESTIMATE. The contrast sums of squares, each with one degree of freedom, are given. Dividing these sums of squares by the error mean square from the analysis of variance table results in *F*-

statistics for testing the null hypotheses that the contrasts have value zero. Associated p-values are computed from F-distributions with 1 degree of freedom in the numerator and error d.f. in the denominator. The F-values resulting from the CONTRAST command are the squares of the corresponding t-values from ESTIMATE, and, of course, the p-values are identical.

The contrast sums of squares are useful in giving an idea of how much of the total amount of variation among the means can be explained by that particular contrast among the means. If a complete set of orthogonal contrasts is written, as in Example 9.2.4, then a complete (although not unique) explanation of the treatment-to-treatment variation can be obtained. In Figures 9.5 and 9.6, the four contrasts excluding 'SvsAH' form a complete orthogonal decomposition of the model sum of squares:

Treatments	$SS[T] = 388.7$	d.f.$= 4$
S vs Others	259.200	1
A vs B	90.250	1
AL vs AH	36.125	1
BL vs BH	3.125	1

From the relative sizes of the contrast sums of squares, we see that there is a big difference between the standard treatment on the one hand, and the experimental treatments, on the other. Furthermore, there is a big difference between those experimental treatments involving drug A and those involving drug B. Finally, while it matters whether or not drug A is at high level, the level of B doesn't seem to have much of an effect. Note that this set of comparisons seems to be a logical set to make; however, it is not the only set that might be of interest, nor is it the only possible complete orthogonal decomposition of SS[T]. Another set of complete, mutually orthogonal contrasts could be written to emphasize comparisons between low and high levels of the experimental drugs, averaging across drugs, for example.

Presumably, the researcher tests or estimates contrasts of interest to the study being conducted. Thus, while a complete set of $k - 1$ mutually orthogonal contrasts can be written which will completely decompose the k-degree-of-freedom model sum of squares into disjoint portions, this is not to imply that the researcher *should* test a complete set of mutually orthogonal contrasts or that one *should not* test contrasts which are not orthogonal.

9.3 Testing Comparisons Suggested by the Data

Example 9.8. Suppose that, upon examining the data in Example 9.1 the investigator noticed that the mean drop in blood pressure for the two treatments involving drug B are higher than those for two treatments involving drug A. Surprised by this unexpected result, the investigator decided to see whether the data indicate a significant difference, at $\alpha = 0.05$ level, between the expected difference in the BP reductions for treatments S + A and S + B.

Significance probabilities reported by SAS assume that the hypotheses were specified prior to the collection and analysis of the data. In order to test hypotheses suggested by the data, the standard error of the contrast from the ESTIMATE statement may be used to compute the Scheffé critical contrast value (CCV) at a desired significance level. The estimated contrast is then compared to the Scheffé CCV.

If the contrast comparing the average response under the two dosage levels of drug A to the average at the two levels of drug B were suggested by the data, instead of being a logical *a priori* hypothesis to test, then the Scheffé CCV is

$$\text{CCV(Scheffé)} = 1.2682\sqrt{4(3.06)}$$

$$= 4.43,$$

where 1.2682 is the estimated standard error of the contrast, found in Figure 9.6, $4 = t - 1$ where t = number of treatments, and 3.06 is the 5% critical value of an F-distribution with 4 and 15 degrees of freedom.

Since the estimate of the contrast, 4.75 (also from Figure 9.6), is larger than the Scheffé CCV, it is declared significant at a = 0.05, even if no such difference was suspected until after the results were observed.

9.4 Multiple Comparison of Means

Example 9.12. Let's use the multiple comparison methods described in this section to test the significance of the four comparisons in Example 9.5, first at the comparisonwise error rate α = 0.05, and then at the experimentwise α = 0.05.

The *p*-value provided for the *F*- or *t*-test of any contrast among treatment means gives the probability, under the null hypothesis, of observing a value of the contrast as favorable to H_1 : as the one observed. That is, the *p*-value can be thought of as a *comparisonwise p-value*. In order to test the four contrasts with a comparisonwise error rate of α = 0.05, then, for two-sided alternative hypotheses, simply compare each of the *p*-values in Figure 9.6 to 0.05, rejecting the null hypothesis of no difference for those contrasts where $p < 0.05$. For one-sided alternative hypotheses, compare $p/2$ to α = 0.05 and reject the null hypothesis of no difference if $p/2 < 0.05$. Using this criterion and assuming one-sided alternatives, the differences between the standard treatment on the one hand and the four experimental conditions on the other, the difference between drugs A and B, and the difference between low and high levels of drug A would be declared significant.

In order to use the Bonferroni method to control the experimentwise error rate at α = 0.05, simply compare each of the *k* printed values of *p* (for two-sided tests) or of $p/2$ (for one-sided tests) to α/k--in this case, 0.05/4 = 0.0125. The Bonferroni criterion declares only the difference between drugs

A and B , and the difference between the standard therapy and the average of the experimental therapies to be significant, and finds no difference in high and low dosage for either drug A or drug B.

The estimates of the contrasts and their standard errors from Figure 9.6 can be used to compute Scheffé CCV's, as in the last section, if it is desired to use this criterion for controlling the experimentwise error rate. Compute

$$\sqrt{(t-1)F(t-1, dfE, \alpha)}$$

times the standard error of the contrast to obtain the Scheffé CCV.

In this example, we compute

$$\sqrt{4F(4,15,0.05)} = \sqrt{4(3.06)} = 3.49857$$

and multiply each of the standard errors by this constant:

SvsOthers	3.49857(1.4179) = 4.9606
ALvsAH	3.49857(1.7935) = 6.2747
BLvsBH	3.49857(1.7935) = 6.2747
AvsB	3.49857(1.2682) = 4.4369

These are the Scheffé contrast critical values. Since the estimate of the contrast SvsOth is 9.0, which is greater than 4.9606, we are able to declare this difference significant. The AvsB comparison, estimated as 4.75, also exceeds the Scheffé CCV = 4.4369 and is therefore declared significant. However, neither ALvsAH nor BLvsBH is found to be significant.

9.5 Multiple Pairwise Comparisons of Treatment Means

Example 9.12. Caceres (1990) used a completely randomized design with six replications per treatment to compare the digestibility, in sheep, of four types of hay: June-harvested Mott elephant grass (JM), September-harvested Mott elephant grass (SM), June-harvested Pensacola bahia grass (JP) and September-harvested Pensacola bahia grass (SP). The experimental animals--24 crossbred mature wethers, each of approximately 60-kg body weight--were randomly allotted to individual metabolism pens. The four types of hay were randomly assigned to six pens each. The measured organic mater content (expressed as a percentage of total dry matter) in the feces of animals fed the four types of hay is as follows... Let's use the six pairwise comparison methods (with $\alpha = 0.05$) to compare the four means.

Reconstructing a Data Set

The first issue to be addressed in using SAS to analyze these data is that we do not have access to the actual data, but rather to only the summary statistics. This summary information is all that is needed to perform pairwise comparisons among treatment means, but unfortunately, the GLM procedure expects to start with raw data, instead of in the middle of the analysis. What we need to do is to "trick" SAS into "thinking" that it has the raw data.

Essentially, what we do is read in a data set with the group means as above and with pooled across-treatment variance equal to 3.20 = MS[E]. You might think that you could spend a lot of time trying to make up such a data set. However, the input statement in the program in Figure 9.7 accomplishes this task relatively easily. "Data" are created which are obtained by adjusting the treatment means (60.3, 63.8, 67.2, and 67.1) up and down a bit, by a function of the within-treatment standard deviations. Consistent with the usual ANOVA assumption, we assume that all groups had the same standard deviation, estimated by the ROOT MSE = 1.7888544.

Figure 9.7. **SAS program to perform pairwise comparisons: data constructed from summary statistics**

```
options ls=70 ps=54;
/*Anova from Summary Statistics*/
data;
  input tmt $ n ybar sd;
    delta = sqrt((n-1)/2)*sd;
    weight = n - 2; y = ybar; output;
    weight = 1; y = ybar - delta; output;
    y = ybar + delta; output;
lines;
JP 6 60.3 1.7888544
SP 6 63.8 1.7888544
JM 6 67.2 1.7888544
SM 6 67.1 1.7888544
;
run;
proc print;
run;
proc glm;
  class tmt;
  freq weight;
  model y = tmt;
  means tmt/lsd scheffe bon tukey snk duncan;
run;
```

Figure 9.8. Artificial data set constructed from summary statistics

OBS	TMT	N	YBAR	SD	DELTA	WEIGHT	Y
1	JP	6	60.3	1.78885	2.82843	4	60.3000
2	JP	6	60.3	1.78885	2.82843	1	57.4716
3	JP	6	60.3	1.78885	2.82843	1	63.1284
4	SP	6	63.8	1.78885	2.82843	4	63.8000
5	SP	6	63.8	1.78885	2.82843	1	60.9716
6	SP	6	63.8	1.78885	2.82843	1	66.6284
7	JM	6	67.2	1.78885	2.82843	4	67.2000
8	JM	6	67.2	1.78885	2.82843	1	64.3716
9	JM	6	67.2	1.78885	2.82843	1	70.0284
10	SM	6	67.1	1.78885	2.82843	4	67.1000
11	SM	6	67.1	1.78885	2.82843	1	64.2716
12	SM	6	67.1	1.78885	2.82843	1	69.9284

This creates numbers whose mean is the same as the group mean and whose standard deviation is the same as the group standard deviation. The interested reader can: work through the algebra indicated and verify the procedure. The "data" set so created is shown in Figure 9.8. The analysis of variance shown in Figure 9.9 verifies that the method works correctly.

Figure 9.9. Analysis of variance produced from artificial data set

```
                    General Linear Models Procedure
                       Class Level Information

                  Class     Levels     Values

                  TMT          4       JM JP SM SP


           Number of observations in data set = 24


Dependent Variable: Y
Frequency:          WEIGHT

Source                  DF     Sum of Squares     F Value       Pr > F

Model                    3       192.84000000       20.09       0.0001

Error                   20        64.00000129

Corrected Total         23       256.84000129

           R-Square                  C.V.                    Y Mean

           0.750818                2.769124               64.6000000
```

Pairwise Comparisons

What is of immediate interest to us in Figure 9.7 are the instructions for performing all pairwise comparisons among the four treatment means. Listed as options in the MEANS statement in GLM (or in ANOVA) are six of the many procedures provided by SAS to perform multiple pairwise comparisons. The default significance level for the pairwise tests is $\alpha = 0.05$. This can be changed by specifying ALPHA=p after the slash in the MEANS command, where p may be any value between 0.0001 and 0.9999 (except restricted to 0.01, 0.05, and 0.10 for the DUNCAN option).

Let us examine the output produced in Figure 9.10 by the (Fisher's) LSD option. We are reminded first in Figure 9.10 that each individual contrast between means is being tested at $\alpha = 0.05$. The least significant difference, along with information needed to compute it, follows. The treatment means are listed in descending order of magnitude, and those means whose difference does not exceed the LSD are bracketed by having a common letter written next to them. (For continuity, the letter is also written on the line between the lines on which the means are printed.) We conclude from Figure 9.10 that treatments JM and SM produce no difference in mean response, but that JM and SM do differ from both JP and SP, and furthermore, that SP and JP differ from each other.

Figure 9.10. All pairwise comparisons using Fisher's LSD method

```
                  T tests (LSD) for variable: Y

      NOTE: This test controls the type I comparisonwise error rate
            not the experimentwise error rate.

                     Alpha= 0.05  df= 20  MSE= 3.2
                        Critical Value of T= 2.09
                  Least Significant Difference= 2.1544

      Means with the same letter are not significantly different.

               T Grouping          Mean        N   TMT

                       A           67.200       6   JM
                       A
                       A           67.100       6   SM

                       B           63.800       6   SP

                       C           60.300       6   JP
```

Output from the other five methods is similar. Figure 9.11 shows the Scheffé pairwise tests. While the least significant difference for the Scheffé tests is considerably larger than that for the Fisher tests, the same conclusions are reached in this example. It is common, however, for different pairs of means to be declared significantly different by different multiple comparison procedures. Running all six of these procedures often results in as much indecision about where differences lie as not performing any tests at all; thus, it is advisable to choose and perform just one as being appropriate for the data at hand. Figures 9.12-9.15 show results from the Bonferroni, Tukey, Student-Newman-Keuls, and Duncan tests, respectively.

Figure 9.11. Scheffé's method for pairwise comparisons

```
                 Scheffe's test for variable: Y

NOTE: This test controls the type I experimentwise error rate
      but generally has a higher type II error rate than REGWF
      for all pairwise comparisons

           Alpha= 0.05  df= 20  MSE= 3.2
           Critical Value of F= 3.09839
           Minimum Significant Difference= 3.1488

Means with the same letter are not significantly different.

      Scheffe Grouping           Mean        N   TMT

                    A           67.200       6   JM
                    A
                    A           67.100       6   SM

                    B           63.800       6   SP

                    C           60.300       6   JP
```

Figure 9.12. Bonferroni pairwise comparisons

```
               Bonferroni (Dunn) T tests for variable: Y

NOTE: This test controls the type I experimentwise error rate,
      but generally has a higher type II error rate than REGWQ.

           Alpha= 0.05  df= 20  MSE= 3.2
           Critical Value of T= 2.93
      Minimum Significant Difference= 3.0231

Means with the same letter are not significantly different.

      Bon Grouping              Mean        N   TMT

                    A          67.200       6   JM
                    A
                    A          67.100       6   SM

                    B          63.800       6   SP

                    C          60.300       6   JP
```

Figure 9.13. Tukey pairwise comparisons

```
        Tukey's Studentized Range (HSD) Test for variable: Y

NOTE: This test controls the type I experimentwise error rate,
      but generally has a higher type II error rate than REGWQ.

            Alpha= 0.05  df= 20  MSE= 3.2
         Critical Value of Studentized Range= 3.958
           Minimum Significant Difference= 2.8907

Means with the same letter are not significantly different.

        Tukey Grouping             Mean      N   TMT

                    A            67.200      6   JM
                    A
                    A            67.100      6   SM

                    B            63.800      6   SP

                    C            60.300      6   JP
```

Figure 9.14. Student-Newman-Keuls pairwise comparisons

```
           Student-Newman-Keuls test for variable: Y

NOTE: This test controls the type I experimentwise error rate
      under the complete null hypothesis but not under partial
      null hypotheses.

            Alpha= 0.05  df= 20  MSE= 3.2

    Number of Means        2         3          4
    Critical Range  2.1543743 2.6129543 2.8907302

Means with the same letter are not significantly different.

        SNK Grouping               Mean      N   TMT

                    A            67.200      6   JM
                    A
                    A            67.100      6   SM

                    B            63.800      6   SP

                    C            60.300      6   JP
```

Figure 9.15. Duncan's pairwise comparisons

```
            Duncan's Multiple Range Test for variable: Y

    NOTE: This test controls the type I comparisonwise error rate,
          not the experimentwise error rate.

              Alpha= 0.05   df= 20   MSE= 3.2

              Number of Means      2      3     4
              Critical Range   2.154  2.261 2.329

    Means with the same letter are not significantly different.

        Duncan Grouping              Mean      N   TMT

                       A            67.200      6   JM
                       A
                       A            67.100      6   SM

                       B            63.800      6   SP

                       C            60.300      6   JP
```

Since in this example there were six observations at each treatment, the least significant difference for comparing treatment means was the same for all pairs of treatments. For unbalanced data, however, the LSD will depend upon which two means are being compared. The output from SAS's pairwise comparison procedures will look different in such cases. Figure 9.16 shows Fisher's LSD tests obtained if the first four values are deleted from treatment JP, for purposes of illustration.

Each pair of treatments is listed in both orders; for example, both JM vs SM and SM vs. JM are given. The estimate of the difference between JM and SM is given as 0.100, while (of course) the difference between SM and JM, in this order, is − 0.100. Giving this estimate the margin for error appropriate to the number of observations in each of the groups, the lower end of the interval comparing JM to SM is − 2.348 and the upper confidence limit is 2.548. Since this interval admits the possibility that the difference between JM and SM could be zero, no significant difference is found between these two means. For comparisons in which both ends of the interval are on the same side of zero, so that the interval does not admit the possibility that the difference between means is zero, the difference is indicated by asterisks.

Figure 9.16. Fisher's LSD tests for unbalanced data

```
                    T tests (LSD) for variable: Y

    NOTE: This test controls the type I comparisonwise error rate
          not the experimentwise error rate.

          Alpha= 0.05  Confidence= 0.95  df= 16  MSE= 4
                    Critical Value of T= 2.11991

 Comparisons significant at the 0.05 level are indicated by '***'.

                         Lower     Difference    Upper
                  TMT    Confidence  Between   Confidence
              Comparison   Limit      Means       Limit

          JM   - SM      -2.348      0.100      2.548
          JM   - SP       0.952      3.400      5.848     ***
          JM   - JP       3.438      6.900     10.362     ***

          SM   - JM      -2.548     -0.100      2.348
          SM   - SP       0.852      3.300      5.748     ***
          SM   - JP       3.338      6.800     10.262     ***

          SP   - JM      -5.848     -3.400     -0.952     ***
          SP   - SM      -5.748     -3.300     -0.852     ***
          SP   - JP       0.038      3.500      6.962     ***

          JP   - JM     -10.362     -6.900     -3.438     ***
          JP   - SM     -10.262     -6.800     -3.338     ***
          JP   - SP      -6.962     -3.500     -0.038     ***
```

It is possible to obtain results in the confidence interval form even for balanced data. To request this type of output, simply write

```
cldiff
```

(for "confidence limits on the differences") after the slash in the MEANS statement.

9.6 Simultaneous Confidence Intervals for Contrasts

Example 9.14.   To determine the effects of four environmental on the growth of a certain species of fish, 12 aquariums in which the water temperature and water movement could be controlled were divided into four groups of three. The groups were randomly assigned to the following four conditions:

1. (SC) Still cold water
2. (SW) Still warm water
3. (FC) Flowing cold water
4. (FW) Flowing warm water

The tanks were stocked with fish of uniform size and fed identical diets over a period of 12 weeks. The mean weights gains (lbs) for a sample of 12 fish obtained by selecting one fish at random from each of the four tanks were as follows:

Treatment	SC	SW	FC	FW
Treatment number i	1	2	3	4
Sample size n_i	3	3	3	3
Treatment mean $\bar{y}_i$	1.55	1.59	1.08	2.01

The corresponding ANOVA table looks like this:

Source	df	SS	MS
Treatments	3	1.302	0.434
Error	8	0.241	0.034

The investigators were mainly interested in estimating the expected differential weight gains:
a) between fish grown in still cold water (SC) and fish grown in flowing cold water (FC);
b) between fish grown in still warm water (SW) and fish grown in flowing warm water (FW);
c) between fish grown in still cold water (SC) and fish grown in still warm water (SW);
d) between fish grown in flowing cold water (FC) and fish grown in flowing warm water (FW).

As in the previous section, we are provided with a summary, instead of with actual data. The program in Figure 9.17 creates a data set with the desired properties and also requests estimates and standard errors of the four contrasts mentioned above. The output from the program appears in Figure 9.18.

Figure 9.17. Program to analyze aquarium data

```
options ls=70 ps=54;
/*Aquarium conditions - Single Factor*/
data;
   input tmt $ n ybar sd;
   delta = sqrt((n - 1)/2)*sd;
   weight = n - 2; y = ybar; output;
   weight = 1; y = ybar - delta; output;
   y = ybar + delta; output;
lines;
SC 3 1.55 0.1735655
SW 3 1.59 0.1735655
FC 3 1.08 0.1735655
FW 3 2.01 0.1735655
;
run;
proc glm;
   class tmt;
   freq weight;
   model y = tmt;
   estimate 'SCvsFC' tmt -1 0 1 0;
   estimate 'SWvsFW' tmt 0 -1 0 1;
   estimate 'SCvsSW' tmt 0 0 1 -1;
   estimate 'FCvsFW' tmt 1 -1 0 0;
run;
```

Simultaneous $100(1-\alpha)$% confidence intervals for the four contrasts can be constructed by multiplying the standard errors of the contrasts (all 0.14 in this case) by the appropriate factor for Scheffé, Bonferroni, or Tukey intervals. Note that since *not all* pairwise differences are being estimated, we would not want to use the MEANS command with the CLDIFF option. The size of the "family" of contrasts to be estimated here is $k = 4$; for all pairwise comparisons among $t = 4$ means, the family size is 6.

Figure 9.18.   Analysis of variance and estimates of contrasts for aquarium data

```
                  General Linear Models Procedure
                     Class Level Information

                 Class     Levels     Values

                 TMT          4       FC FW SC SW


            Number of observations in data set = 12


Dependent Variable: Y
Frequency:          WEIGHT

Source                  DF     Sum of Squares     F Value     Pr > F

Model                    3        1.30162500        14.40     0.0014

Error                    8        0.24099986

Corrected Total         11        1.54262486

                 R-Square              C.V.                  Y Mean

                 0.843773           11.14385            1.55750000


Source                  DF        Type I SS     F Value     Pr > F

TMT                      3        1.30162500        14.40     0.0014

Source                  DF       Type III SS     F Value     Pr > F

TMT                      3        1.30162500        14.40     0.0014

                                T for H0:     Pr > |T|     Std Error of
Parameter         Estimate      Parameter=0                  Estimate

SCvsFC           0.47000000          3.32       0.0106      0.14171564
SWvsFW          -0.42000000         -2.96       0.0180      0.14171564
SCvsSW          -0.04000000         -0.28       0.7849      0.14171564
FCvsFW          -0.93000000         -6.56       0.0002      0.14171564
```

The factor for computing the margin for error for 95% simultaneous confidence intervals using the Scheffé method is

$$\sqrt{(t-1)F(t-1,\nu,\alpha)} = \sqrt{3(4.07)} = 3.4943,$$

the factor for the Bonferroni intervals is

$$t(\nu, \alpha / 2k) = 3.206,$$

and the one for the Tukey intervals is

$$2^{-1/2} q(t, \nu, \alpha) = 2^{-1/2} (4.53) = 3.203.$$

9.7 Predicting Linear Functions of Sample Means

Example 9.16. In Example 9.14 we considered a study designed to compare the mean weight gain of fish grown in four environmental conditions: SC (still cold water), SW (still warm water), FC (flowing cold water), and FW (flowing warm water). Let Y_{SC}, Y_{SW}, Y_{FC}, and Y_{FW} denote the gains in weight of four fish grown under the four conditions SC, SW, FC, and FW, respectively. Let's construct a set of 95% simultaneous prediction intervals for the following four comparisons of the weight gains.

$$\hat{\theta}_{1f} = Y_{SC} - Y_{FC}$$
$$\hat{\theta}_{2f} = Y_{SW} - Y_{FW}$$
$$\hat{\theta}_{3f} = Y_{SC} - Y_{SW}$$
$$\hat{\theta}_{4f} = Y_{FC} - Y_{FW}$$

We can regard Y_{SC}, Y_{SW}, Y_{FC}, and Y_{FW} as the means of four future random samples, each of size $m = 1$ from the normally distributed populations from which the data in Example 9.14 have been collected. Then the problem can be stated as the problem of constructing simultaneous prediction intervals for $k = 4$ linear functions of the means of four future random samples of sizes $m_1 = m_2 = m_3 = m_4 = 1$.

In each of the comparisons above, the coefficients for the linear functions of the means include one +1 and one −1. Thus, for each of the four comparisons, the margin for error will be the same. For simultaneous intervals, we have

$$t(N - t, 0.05, 4) = t(8, 0.05, 4) = 3.206,$$

MS[E] = 0.2410/8 = 0.0301 from Figure 9.18, and

$$\sqrt{[c_i^2 (\frac{1}{m_i} + \frac{1}{n_i}) + c_j^2 (\frac{1}{m_j} + \frac{1}{n_j})] MS[E]}$$

$$= \sqrt{[(1)^2 (\frac{1}{1} + \frac{1}{3}) + (-1)^2 (\frac{1}{1} + \frac{1}{3})](0.0301)}$$

$$= 0.2833$$

Then the simultaneous prediction intervals are formed by adding and subtracting

$$3.206(0.2833) = 0.91$$

to/from the estimated value of the contrast. (Slight differences in these results and Rao's are due to rounding error.)

9.8 Determining Sample Sizes

Example 9.17. On the basis of a careful literature review and some preliminary studies, an investigator expects that the weight gains (kg) of sheep fed three experimental diets will be in the neighborhood of $\mu_1 = 13$, $\mu_2 = 16$ and $\mu_3 = 22$. To confirm this hypothesis, the investigator wants to conduct a completely randomized experiment in which n animals will be allocated to each of the three diets and the weight gain after 12 weeks on the diet will be measured for each animal. The experimenter wants to ensure that, if hisresearch hypothesis is true, there will be at least an 80%

chance that a 0.05-level F-test will reject the null hypothesis H_0: $\mu_1 = \mu_2 = \mu_3$. What should the value of n be?

Example 9.18. Suppose that the researcher in Example 9.17 conservatively estimates the standard deviation of the weight gains of animals on any of the experimental diets to be about $\sigma = 5$ kg. Let's determine the sample size such that, if $\sigma_\mu^2 = 14$, the probability that H_0: $\sigma_\mu^2 = 0$ will be rejected at the $\alpha = 0.05$ level is 80%.

O'Brien's FPOWTAB program, (which was introduced in Chapter 7 and is given in the Appendix to Chapter 7) can be used to determine sample sizes for comparing means. Figure 9.19 shows the specification file corresponding to Example 9.17. Record 1 gives the title of the exercise as "Rao Example Comparing Three Animal Diets." Record 2 says that there are three means to be compared, and if samples of size 1 were taken under each diet, then a total "basis" sample size would be 3. Record 3 says that there is only one scenario of hypothesized mean differences being considered (namely, $\mu_1 = 13$, $\mu_2 = 16$ and $\mu_3 = 22$), and this scenario is titled, "Hypothesized Effect." Record 4 says that there is only one significance level being considered, namely $\alpha = 0.05$, and Record 5 says that there is only one value for the standard deviation being proposed, namely $\sigma = 5$. Four group sizes, $n = 6, 7, 8$, and 10, are specified in Record 6. Finally, Record 7 assigns the title "Diet Effect" to the test of H_0: $\mu_1 = \mu_2 = \mu_3$. Since there are three means to be compared the numerator degrees of freedom for the F-test will be $3 - 1 = 2$. The value for SSH(Population) is

$$\Sigma(\mu_i - \overline{\mu})^2,$$

which in this example is

$$(13 - 17)^2 + (16 - 17)^2 + (22 - 17)^2 = 42$$

since the average of the three group means is

$$(13 + 16 + 22)/3 = 17.$$

Figure 9.19. FPOWTAB specification file for Examples 9.17 and 9.18

```
Rao Example Comparing Three Animal Diets
3 3
1 Hypothesized Effect
1 0.05
1 5
4 6 7 8 10
Diet Effect   2 42
```

Figure 9.20 shows the power analysis for these specifications. Consistent with Rao's conclusion in Example 9.18, we see that 21 animals all told, or $n = 7$ animals per diet will be sufficient to detect the hypothesized effect with at least 80% certainty.

Figure 9.20. Power analysis for animal diet example

```
              Rao Example Comparing Three Animal Diets

       Effect: Diet Effect,
       Degrees of Freedom Hypothesis: 2,
       Scenario: Hypothesized Effect,
       Powers Computed from SSH(Population): 42,
       Using the Basis Total Sample Size: 3,
       AND Total Cells in Design: 3
       -----------------------------------------
       |                           |   Std. Dev   |
       |                           |--------------|
       |                           |      5       |
       |                           |--------------|
       |                           |   Total N    |
       |                           |--------------|
       |                           |18 |21 |24 |30 |
       |                           |---+---+---+---|
       |                           |PO-|PO-|PO-|PO-|
       |                           |WER|WER|WER|WER|
       |---------------------------+---+---+---+---|
       |Test Type  |ALPHA          |   |   |   |   |
       |-----------+-----------|    |   |   |   |   |
       |Regular F  |0.05           |.73|.81|.87|.94|
       -----------------------------------------
```

Example 9.19. Suppose the investigator in Example 9.17 wants 80% power for a 0.05-level ANOVA F-test if the differential weight gain between at least one pair of diets exceeds 8 kg. Determine the appropriate sample size if $\sigma^2 = 25.0$ (kg)^2.

The only modification necessary to the specification file in Figure 9.19 is that instead of entering SSH(Population) = 42 in Record 7, use

$$\text{SSH(Population)} = \Delta^2/2,$$

where Δ is the difference in means that is to be detected; in this case, 8. Replacing the 42 by 32 in Figure 9.19, and trying sample sizes $n = 7, 8, 9$, and 10 produces Figure 9.21. Consistent with Rao's conclusion in Example 9.19, we see that using $n = 9$ animals per diet will achieve the desired results.

Figure 9.21. Power analysis for Example 9.19

```
Second Rao Example Comparing Three Animal Diets

    Effect: Diet Effect,
    Degrees of Freedom Hypothesis: 2,
    Scenario: Hypothesized Effect,
    Powers Computed from SSH(Population): 32,
    Using the Basis Total Sample Size: 3,
    AND Total Cells in Design: 3
-------------------------------------------
|                       |    Std. Dev    |
|                       |----------------|
|                       |        5       |
|                       |----------------|
|                       |     Total N    |
|                       |----------------|
|                       |21  |24  |27  |30 |
|                       |---+---+---+---|
|                       |PO-|PO-|PO-|PO-|
|                       |WER|WER|WER|WER|
|-----------------------+---+---+---+---|
|Test Type   |ALPHA     |   |   |   |   |
|------------+----------|   |   |   |   |
|Regular F   |0.05      |.69|.76|.82|.87|
-------------------------------------------
```

Example 9.20. Suppose we are interested in planning a completely randomized experiment to study the difference between the response times to four methods of treating migraine headache. An objective in such a study might be to test the null hypothesis that there is no difference between the mean response times against the research hypothesis that the means do differ. We regard the differences between the four treatments as practically important if, for at least one pair of treatments, the probability that response time under one of the treatments will be at least as large as that under the other is 90%. Let's determine the sample size necessary to guarantee that the 0.05-level ANOVA F-test will have 80% power to detect practically important differences between the four treatments.

In order to use FPOWTAB to determine sample size in this problem, some algebra shows that the quantity

$$\frac{\Delta}{\sigma\sqrt{2}}$$

in Rao's Example 9.20 corresponds to

$$\frac{\sqrt{SSH(Pop)}}{\sigma}.$$

Since the normal curve ordinate corresponding to a cumulative normal probability of 0.90 is $z(0.10) = 1.282$, we have

$$\frac{\sqrt{SSH(Pop)}}{\sigma} = 1.282$$

or

$$\sqrt{SSH(Pop)} = 1.282\sigma.$$

We can arbitrarily assign the value 1 to σ and use

$$SSH(Pop) = (1.282)^2 = 1.6435$$

in Record 7 in the specification file. Figure 9.22 shows the specification file for this example, and the FPOWTAB output is given in Figure 9.23. Power

Figure 9.22. Specification file for Example 9.20

```
Four Crop Conditions
4 4
1 P(C1>=C2)=.9
1 0.05
1 1
5 4 5 6 8 10
Conditions   3 1.6435
```

Figure 9.23. Power analysis for Example 9.20

```
                    Four Crop Conditions

        Effect: Conditions,
        Degrees of Freedom Hypothesis: 3,
        Scenario: P(C1>=C2)=.9,
        Powers Computed from SSH(Population): 1.6435,
        Using the Basis Total Sample Size: 4,
        AND Total Cells in Design: 4
        --------------------------------------------------
        |                         |    Std. Dev         |
        |                         |---------------------|
        |                         |          1          |
        |                         |---------------------|
        |                         |     Total N         |
        |                         |---------------------|
        |                         |16 |20 |24 |32 |40 |
        |                         |---+---+---+---+---|
        |                         |PO-|PO-|PO-|PO-|PO-|
        |                         |WER|WER|WER|WER|WER|
        |-------------------------+---+---+---+---+---|
        |Test Type   |ALPHA       |   |   |   |   |   |
        |------------+------------|   |   |   |   |   |
        |Regular F   |0.05        |.42|.55|.66|.82|.91|
        --------------------------------------------------
```

of 0.82 is attained by making $n = 8$ observations in each of the four conditions.

Before closing, we note that when values for the population means are hypothesized, the GLM procedure can be used to compute SSH(Population) values for use in the specification files. The program in Figure 9.24 illustrates using GLM procedure to compute SSH(Population) for the scenario in Example 9.17. Recall that in that example, it was hypothesized that the three means would be in the neighborhoods of 3, 6, and 12, respectively.

Figure 9.24. Program to use GLM to compute SSH(Population)

```
options ls=70 ps=54 pageno=1;
title 'Get SSH(Pop) Values for Rao Ex.
       9.17';
data;
  input diet scnario1 basen;
cards;
1   3 1
2 6 1
3 12 1
;
run;
proc glm;
   class diet;
   freq basen;
   model scnario1=diet/ss3;
run;
```

The program in Figure 9.24 can be modified easily to allow for exploration of the effects of different scenarios and can be used in instances in which unbalanced designs are required.

Figure 9.25 shows the output from the program. Note that since only one "data" value per group was inputted, there is no way to evaluate error. However, that is of no importance to us here: all we want is the treatment sum of squares and degrees of freedom. These are found as the Model Sum of Squares and degrees of freedom, and also under the Type III SS and d.f. Since only the hypothesis of equality of treatment means was to be tested, the REG procedure could have been used. The treatment sum of squares would have been found as the Model Sum of Squares. However, in multifactor studies, as will be seen in Chapter 13, the Model Sum of Squares and degrees of freedom will reflect the combined influences of all factors, and using the GLM procedure will produce the Type III Sums of Squares and degrees of freedom which decompose the Model SS into its component parts.

Figure 9.25. GLM computes SSH(Population)

```
                    Get SSH(Pop) Values for Rao Ex. 9.17

                      General Linear Models Procedure
                           Class Level Information

                    Class      Levels     Values

                    DIET          3       1 2 3

               Number of observations in data set = 3

Dependent Variable: SCNARIO1
Frequency:          BASEN

Source                    DF      Sum of Squares    F Value       Pr > F

Model                      2        42.00000000        .            .

Error                      0             .

Corrected Total            2        42.00000000

                R-Square                  C.V.           SCNARIO1 Mean

                1.000000                    0              7.00000000


Source                    DF      Type III SS       F Value       Pr > F

DIET                       2        42.00000000        .            .
```

10

Simple Linear Regression

10.1 Introduction

Either the *General Linear Model* (GLM) procedure or the *Regression* (REG) procedure may be used to perform a simple linear regression analysis, even though both procedures are designed to handle much more complex analyses. REG may be used for simple and multiple regression analyses, and GLM can perform not only simple and multiple regression analyses but also can analyze a large variety of other linear models, as we saw in Chapters 8 and 9. The LOGISTIC procedure is used when the response variable is dichotomous or ordinal.

10.2 Estimating Parameters of the Simple Linear Regression Model

Example 10.7. To determine if the age of a rat (dependent variable) can be predicted using independent variables such as body weight, hematocrit (percentage of red blood cells), and protein (mg/ml blood), a study was conducted on a total of 37 rats ranging in age from 20 to 89 days. The data on two variables--Y = age and X = weight--for $n = 10$ rats are as follows... Let's suppose that a simple linear regression model is appropriate for these data.

Figure 10.1 shows a SAS program using the REGression procedure to estimate the regression parameters and to plot the data and the regression equation.

Figure 10.1. **Using the REG procedure to estimate regression parameters**

```
options ls=70 ps=54;
/*Simple Regression*/
data rats;
  input weight age @@;
lines;
76 28 89 33 154 35 189 37 180 38
180 47 241 66 320 67 271 70 370 82
;run;
proc reg;
  model age = weight/p;
  id weight;
  plot age * weight p. * weight = '*'/
      overlay;
run;
```

Just as was the case when using the ANOVA or the GLM procedure, a MODEL statement is required to indicate the dependent and independent variables. Note that no CLASS statement is used in REG, however, since the independent variables in a regression are numerical, rather than categorical. Following the slash in the model statement, writing the letter "p" requests that the predicted values and residuals corresponding to each case in the data set be printed. This option is requested here for purposes of instruction and is probably not often requested in actual applications, especially if the data set is large. The ID WEIGHT statement asks that each of the cases in the printout of the predicted values and residuals be identified by the weight (*X*-value) of the rat.

The regression procedure has a built-in plotting facility which enables us to plot the data and the estimated regression line on a single graph, using the OVERLAY option. The plot of AGE by WEIGHT graphs the data points, and the plot of predicted values by WEIGHT produces points on the regression line which can then be connected. Note that the predicted values are denoted by

p.

and it is necessary to include the period after the symbol "p." To distinguish predicted *Y*-values from actual ones, predicted values are plotted with an asterisk (*).

In some applications, it might be convenient to output the predicted values to a data set so that they can be used in subsequent computations. Writing

```
output out=pred p=yhat;
```

will create a new data set named "pred" which contains the *X*- and *Y*-values in the data set "rats" and also the predicted values, under the name "yhat". Note that the names in italics in the above output statement are names supplied by the programmer. Also note that in outputting the predicted values to a data set, it is not necessary to follow the symbol p with a period. If the predicted values are outputted to a data set, then the plot of data with predicted values can be constructed using the PLOT procedure, as in

```
proc plot data=pred;
       plot age*weight yhat*weight='*'/overlay;
```

Figure 10.2 shows the output from the regression procedure, exclusive of the plot of the data and predicted values. There is a lot of information given, all of which will be explained eventually. Note that there are three sections to the printout in Figure 10.2. First is an analysis of variance table, similar to the ANOVA table produced by the ANOVA or GLM procedure, along with associated statistics such as R-Square. Then there is a section labeled "Parameter Estimates," and finally, there is a section listing the *X*- and *Y*-values, predicted values, and residuals for each rat.

Refer to the section of the output labeled "Parameter Estimates." Under the heading "Variable" are listed INTERCEP(T) and WEIGHT, and under the heading "Parameter Estimate" are given estimates of the regression intercept and the coefficient on the WEIGHT variable, respectively. That is

Figure 10.2. Regression analysis for rat age and weight data

```
Model: MODEL1
Dependent Variable: AGE


                        Analysis of Variance

                      Sum of         Mean
Source          DF    Squares       Square      F Value    Prob>F

Model           1   2909.95053   2909.95053      61.562    0.0001
Error           8    378.14947     47.26868
C Total         9   3288.10000


     Root MSE        6.87522    R-square     0.8850
     Dep Mean       50.30000    Adj R-sq     0.8706
     C.V.           13.66843

                      Parameter Estimates

                 Parameter      Standard     T for H0:
Variable   DF     Estimate        Error    Parameter=0    Prob > |T|

INTERCEP    1    10.886288    5.47363354        1.989       0.0819
WEIGHT      1     0.190404    0.02426727        7.846       0.0001

                               Dep Var    Predict
                 Obs   WEIGHT     AGE       Value    Residual

                   1       76   28.0000   25.3570     2.6430
                   2       89   33.0000   27.8323     5.1677
                   3      154   35.0000   40.2086    -5.2086
                   4      189   37.0000   46.8727    -9.8727
                   5      180   38.0000   45.1591    -7.1591
                   6      180   47.0000   45.1591     1.8409
                   7      241   66.0000   56.7737     9.2263
                   8      320   67.0000   71.8157    -4.8157
                   9      271   70.0000   62.4859     7.5141
                  10      370   82.0000   81.3359     0.6641

Sum of Residuals                       0
Sum of Squared Residuals         378.1495
Predicted Resid SS (Press)       522.4303
```

$$\hat{\beta}_0 = 10.886288$$

and

$$\hat{\beta}_1 = 0.190404.$$

Substituting the sample values of X = weight into the equation

$$\hat{y} = 10.886288 + 0.190404x$$

produces the predicted values in Figure 10.2. Compare these values to Rao's in Section 10.3.   Subtracting predicted Y-values from actual Y-values produces the residuals in Figure 10.2. Note that some residuals are positive and others negative so that their sum is zero.   The sum of the squared residuals is

$$SS[E] = 378.1495,$$

which is the same as Rao gets, apart from rounding error, in Section 10.3. Dividing the error sum of squares by its degrees of freedom gives the error mean square,

$$MS[E] = 47.26868,$$

which can be found in the analysis of variance table at the top of Figure 10.2 and in Rao's Example 10.8. The Predicted Residual Sum of Squares (Press) will be explained later.

Plotting the (X, Y)-values on the same graph with the $(X, \hat{Y})$-values, with these latter indicated by asterisks, produces Figure 10.3. Apart from the limitations of the printer, the asterisks all lie in a straight line, which may be drawn in by hand. Compare this figure to Rao's Figure 10.7. The question mark in Figure 10.3 indicates that a predicted value is so close to an actual Y-value that the printer cannot print them with two separate symbols.

Figure 10.3. Graph of data with predicted values superimposed

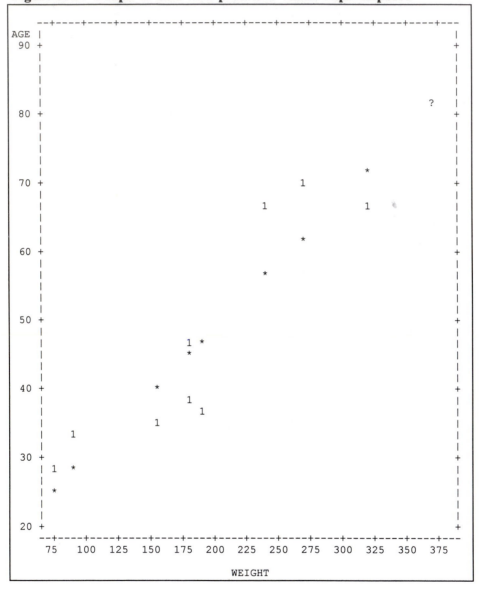

10.3 Inferences about Population Means and Predicting Future Values

Example 10.9. To understand the pattern of variation in plant species in Mediterranean grasslands, data were collected on Y = density of plant species (the number of species per 0.04 m^2) and X = the altitude of the region (in 1000 m) from 12 experimental plots over a period of five years (1986 to 1990). The results obtained for the average (over the 5-year period) species density are as follows...

Figure 10.4 is a SAS program to regress species density on altitude and obtain and graph confidence intervals for expected responses. The parameter estimates and analysis of variance table are shown in Figure 10.5.

Figure 10.4. SAS program regressing species density on altitude

```
options ls=70 ps=54;
/*Species density vs altitude*/
data;
  input alt @@;
    do i = 1 to 2;
      input density @@;
      output;
    end;
  drop i;
lines;
0.64 15   20
0.86 16   18.5
0.89 13.5 16
1.22 11   19.5
1.45 11.5 12
1.72 8    8.5
;
run;
proc reg;
  model density = alt/clm;
  plot density*alt (u95m. l95m. p.) *alt='*'/
      overlay;
  id alt;
run;
```

Figure 10.5. Analysis of variance for plant density data

```
Model: MODEL1
Dependent Variable: DENSITY


                          Analysis of Variance

                        Sum of          Mean
Source          DF      Squares         Square      F Value      Prob>F

Model            1     110.62893      110.62893      15.707       0.0027
Error           10      70.43357        7.04336
C Total         11     181.06250


      Root MSE        2.65393      R-square        0.6110
      Dep Mean       14.12500      Adj R-sq        0.5721
      C.V.           18.78890


                         Parameter Estimates

                    Parameter      Standard     T for H0:
      Variable  DF   Estimate        Error     Parameter=0     Prob > |T|

      INTERCEP   1   23.354287    2.45153872       9.526         0.0001
      ALT        1   -8.167511    2.06084435      -3.963         0.0027
```

The parameter estimates,

$$\hat{\beta}_0 = 23.354287$$

and

$$\hat{\beta}_1 = -8.167511$$

may be compared with Rao's results in Section 10.4.

Next to the parameter estimates in Figure 10.5 are given estimates of their respective standard errors, which can be used to construct confidence interval estimates of the regression parameters and to test hypotheses concerning them. The estimated standard error of $\hat{\beta}_0$ is given as

$$\hat{\sigma}_{\hat{\beta}_0} = 2.45153872$$

and the estimated standard error of $\hat{\beta}_1$ is

$$\hat{\sigma}_{\hat{\beta}_1} = 2.06084435,$$

as Rao obtains in Section 10.4. Then a 95% confidence interval for the slope parameter is

$$-8.1675 \pm (2.228)(2.0608)$$

or

$$-12.7590 < \hat{\beta}_1 < -3.5760$$

since $t(10, 0.025) = 2.228$. (Recall that the degrees of freedom for MS[E] are $n - 2$ for a simple linear regression.) A confidence interval for $\hat{\beta}_0$ could be constructed in a similar fashion if the intercept were interpretable in this example.

Values of t-statistics for testing

$$H_0 : \beta_0 = 0$$

and

$$H_0 : \beta_1 = 0$$

are provided automatically by SAS in the column labeled

```
         T for H0
     Parameter = 0
```

In Figure 10.5 we see that to test whether the intercept parameter differs from zero, the value of the t-statistic is

$$t = 9.526$$

and the t-value for testing the slope parameter is

$$t = -3.963.$$

Additionally, under the heading

$$\text{Prob} > |T|$$

are given the *p*-values for testing the *two-sided* alternative hypotheses,

$$H_1 : \beta_0 \neq 0$$

and

$$H_1 : \beta_1 \neq 0,$$

respectively. Clearly, for these data we would reject the hypotheses that the intercept and slope parameters are zero.

Refer to the model statement in Figure 10.4. As an option after the slash in the model statement is the command

$$\text{clm}$$

which stands for "*c*onfidence *l*imits on the *m*ean response." This option can be used to compute and print 95% confidence limits on

$$\mu(x) = \beta_0 + \beta_1(x)$$

for every value of X in the sample data. (Note: Only 95% limits are available in REG.) Figure 10.6 shows the printout of the predicted values, residuals and confidence limits on the mean response at each X-value in the sample. For example, mean species density at altitude 640 meters, or for x_0 = 0.64, is estimated to be 18.1271 (under "Predict Value"), with an estimated standard error of 1.268 (under "Std Err Predict). The lower limit of the 95% confidence interval for the mean species density at this altitude is then found under "Lower 95% Mean" to be (apart from rounding error)

Figure 10.6.   Predicted values and confidence limits on the mean response for plant density data

Obs	ALT	Dep Var DENSITY	Predict Value	Std Err Predict	Lower95% Mean	Upper95% Mean
1	0.64	15.0000	18.1271	1.268	15.3028	20.9513
2	0.64	20.0000	18.1271	1.268	15.3028	20.9513
3	0.86	16.0000	16.3302	0.947	14.2205	18.4400
4	0.86	18.5000	16.3302	0.947	14.2205	18.4400
5	0.89	13.5000	16.0852	0.912	14.0533	18.1171
6	0.89	16.0000	16.0852	0.912	14.0533	18.1171
7	1.22	11.0000	13.3899	0.788	11.6336	15.1463
8	1.22	19.5000	13.3899	0.788	11.6336	15.1463
9	1.45	11.5000	11.5114	1.011	9.2590	13.7637
10	1.45	12.0000	11.5114	1.011	9.2590	13.7637
11	1.72	8.0000	9.3062	1.437	6.1040	12.5083
12	1.72	8.5000	9.3062	1.437	6.1040	12.5083

Obs	ALT	Residual
1	0.64	-3.1271
2	0.64	1.8729
3	0.86	-0.3302
4	0.86	2.1698
5	0.89	-2.5852
6	0.89	-0.0852
7	1.22	-2.3899
8	1.22	6.1101
9	1.45	-0.0114
10	1.45	0.4886
11	1.72	-1.3062
12	1.72	-0.8062

Sum of Residuals		0
Sum of Squared Residuals		70.4336
Predicted Resid SS (Press)		94.0466

$$18.1271 - (2.228)(1.268) = 15.3028$$

and the upper end of the interval is similarly

$$18.1271 + (2.228)(1.268) = 20.9513.$$

Figure 10.7. Graph of confidence limits for the mean response

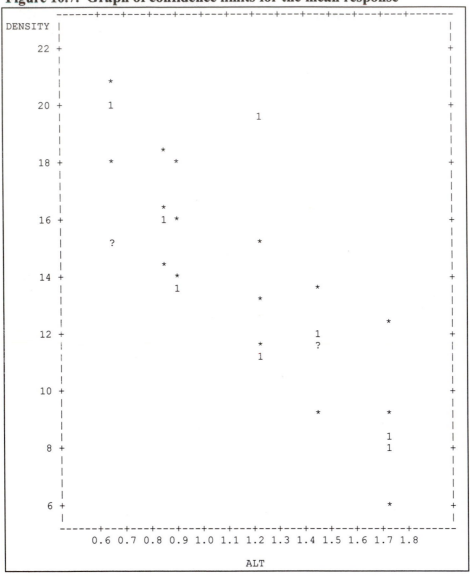

Figure 10.7 shows the data, the predicted values, and the upper and lower confidence limits all plotted on the same graph. Data points are indicated by the number "1", and predicted values and confidence limits are graphed with asterisks (*). The plot command in the REG procedure in Figure 10.4 refers to the upper confidence limits and lower confidence limits as

    u95m.

and

    195m.,

respectively. Enclosing u95m., 195m., and p. in parentheses requests that they all be printed as asterisks, but different symbols could have been used for each. Connecting the predicted values in Figure 10.7 draws a straight line. Connecting the upper limits results in a curve concave from above, and connecting the lower limits forms a curve concave from below, so that the confidence limits are seen to form a curved band. The width of the band is narrowest over $x = \bar{x} = 1.13$.

Example 10.10. On the basis of the plant species data in Example 10.9, let's estimate the mean species density at an altitude of 750 m.

Even though we have no data at an altitude of 750 meters, since this altitude is in the range of altitudes covered by the data, it is reasonable to use the regression equation to make such an estimate. How can we get SAS to compute an estimated mean response and associated confidence limits for an X-value that is not a data value? Figure 10.8 shows how the program from Figure 10.4 can be modified to do this. Changes to the previous program are indicated in boldface. The actual data are entered in a data set called "real." In another data set called "proposed," the value 0.75 for altitude, and a missing value, indicated by a single period, for density is entered. The data set "both" combines the 12 actual data values with the "pseudo" data value and the regression procedure is performed on this combined "data" set.

Figure 10.8. Regression program modified to include a "pseudo" data value

```
options ls=70 ps=54;
/*Species density vs altitude, with "pseudo"
       data value*/
data real;
  input alt @@;
    do i = 1 to 2;
      input density @@;
      output;
    end;
  drop i;
lines;
0.64 15    20
0.86 16    18.5
0.89 13.5 16
1.22 11    19.5
1.45 11.5 12
1.72  8     8.5
;
run;
data proposed;
input alt density;
lines;
0.75 .
;
run;
data both; set real proposed;
run;
proc reg data=both;
  model density = alt/clm;
  plot density*alt (u95m. l95m. p.)
      *alt='*'/overlay;
  id alt;
run;
```

SAS must drop from use in estimating the regression parameters any cases with missing data, so that the parameter estimates and analysis of variance corresponding to the program in Figure 10.8 are identical to those in Figure 10.5. However, SAS will compute the predicted value and confidence limits for the case in which $x_0 = 0.75$, as shown in Figure 10.9.

Figure 10.9. Predicted values and confidence limits at actual and proposed *X*-values

Obs	ALT	Dep Var DENSITY	Predict Value	Std Err Predict	Lower95% Mean	Upper95% Mean
1	0.64	15.0000	18.1271	1.268	15.3028	20.9513
2	0.64	20.0000	18.1271	1.268	15.3028	20.9513
3	0.86	16.0000	16.3302	0.947	14.2205	18.4400
4	0.86	18.5000	16.3302	0.947	14.2205	18.4400
5	0.89	13.5000	16.0852	0.912	14.0533	18.1171
6	0.89	16.0000	16.0852	0.912	14.0533	18.1171
7	1.22	11.0000	13.3899	0.788	11.6336	15.1463
8	1.22	19.5000	13.3899	0.788	11.6336	15.1463
9	1.45	11.5000	11.5114	1.011	9.2590	13.7637
10	1.45	12.0000	11.5114	1.011	9.2590	13.7637
11	1.72	8.0000	9.3062	1.437	6.1040	12.5083
12	1.72	8.5000	9.3062	1.437	6.1040	12.5083
13	0.75	.	17.2287	1.096	14.7876	19.6697

Obs	ALT	Residual
1	0.64	-3.1271
2	0.64	1.8729
3	0.86	-0.3302
4	0.86	2.1698
5	0.89	-2.5852
6	0.89	-0.0852
7	1.22	-2.3899
8	1.22	6.1101
9	1.45	-0.0114
10	1.45	0.4886
11	1.72	-1.3062
12	1.72	-0.8062
13	0.75	.

Sum of Residuals		0
Sum of Squared Residuals		70.4336
Predicted Resid SS (Press)		94.0466

At the line for the 13th observation are the predicted value, 17.23, and the lower and upper 95% confidence limits, 14.79 and 19.67, as in Rao's Example 10.10.

It is a good idea to keep actual data in a separate data set from the SAS data set containing additional values of *X* for which predicted values are required. This is because SAS will always use as much of the data as it can, and while it will have to delete cases with missing values when performing any *bivariate* analyses, SAS will use all available data values when performing *univariate* analyses.   As we have seen, when the REGression procedure was run on the data set "both," only the 12 actual data cases could be used in the computations.   However, if the MEANS procedure were performed on the "both" data set, then the statistics would be computed using the 12 values of density, but statistics for the variable "alt" would be calculated using the *13* available values, one of which was not an actual data value.

> **Example 10.11.** For a particular variety of plant, researchers wanted to . develop a formula for predicting the quantity of seeds as a function of the density of plants. They conducted a study with four levels of the factor X, the number of plants per plot, in a completely randomized design; there were four4 replications per treatment.

In Example 10.11, Rao notes that the methods of Chapter 9 can be used to construct prediction intervals when $X = 10$, 20, 30, or 40—the values used in the study—but that a prediction formula that is valid for values of X not included in the study requires the assumption that the seed yield data satisfy a simple linear regression model. Then, in Example 10.12, Rao gives a 95% prediction interval for the yield of a plot containing 35 plants. SAS can be instructed to compute prediction intervals for a single new observation in much the same way that confidence intervals for the mean responses at various X-values were constructed. Figure 10.10 shows that in order to compute and graph prediction limits, the option CLI (*c*onfidence *l*imits on an *i*ndividual response) is used in place of CLM in the model statement, and the upper and lower limits are referred to as

```
u95. and 195.,
```

respectively. For variety and for ease of reading the graph, predicted values are plotted with the symbol "P" and upper and lower limits are plotted with the symbols "U" and "L", respectively. The analysis of variance and parameter estimates are shown in Figure 10.11, the predicted values and

Figure 10.10. Program computing prediction limits for a single future observation

```
options ls=70 ps=54;
/*Seed yield as a function of plant density*/
data a;
  input density @@;
    do i = 1 to 4;
       input grams @@;
       output;
    end;
  drop i;
lines;
10 12.6 11.0 12.1 10.9
20 15.3 16.1 14.9 15.6
30 17.9 18.3 18.6 17.8
40 19.2 19.6 18.9 20.0
;
run;
data b;
  input density grams;
lines;
35 .
;
run;
data c; set a b;
run;
proc reg data = c;
  model grams=density/cli;
  id density;
  plot u95.*density='U' l95.*density='L'
     p.*density='P' grams*density/overlay;
run;
```

prediction limits are given in Figure 10.12, and the graph is in Figure 10.13. Compare the analysis of variance table in Figure 10.11 to the table in Rao's Example 10.13.

Note that the confidence and prediction intervals given in Figures 10.6 and 10.12 are one-at-a-time intervals. SAS has no provision for computation of Working-Hotelling or Bonferroni simultaneous intervals.

Figure 10.11. Analysis of variance and parameter estimates for plant density data

```
Model: MODEL1
Dependent Variable: GRAMS


                         Analysis of Variance

                       Sum of          Mean
Source           DF    Squares        Square      F Value     Prob>F

Model             1   135.20000     135.20000     180.783     0.0001
Error            14    10.47000       0.74786
C Total          15   145.67000


      Root MSE          0.86479      R-square      0.9281
      Dep Mean         16.17500      Adj R-sq      0.9230
      C.V.              5.34644


                         Parameter Estimates

                    Parameter      Standard     T for H0:
   Variable   DF     Estimate         Error    Parameter=0     Prob > |T|

   INTERCEP    1     9.675000    0.52957193        18.269         0.0001
   DENSITY     1     0.260000    0.01933723        13.446         0.0001
```

Figure 10.12. Predicted values and prediction limits for plant density data

Obs	DENSITY	Dep Var GRAMS	Predict Value	Std Err Predict	Lower95% Predict	Upper95% Predict
1	10	12.6000	12.2750	0.362	10.2645	14.2855
2	10	11.0000	12.2750	0.362	10.2645	14.2855
3	10	12.1000	12.2750	0.362	10.2645	14.2855
4	10	10.9000	12.2750	0.362	10.2645	14.2855
5	20	15.3000	14.8750	0.237	12.9519	16.7981
6	20	16.1000	14.8750	0.237	12.9519	16.7981
7	20	14.9000	14.8750	0.237	12.9519	16.7981
8	20	15.6000	14.8750	0.237	12.9519	16.7981
9	30	17.9000	17.4750	0.237	15.5519	19.3981
10	30	18.3000	17.4750	0.237	15.5519	19.3981
11	30	18.6000	17.4750	0.237	15.5519	19.3981
12	30	17.8000	17.4750	0.237	15.5519	19.3981
13	40	19.2000	20.0750	0.362	18.0645	22.0855
14	40	19.6000	20.0750	0.362	18.0645	22.0855
15	40	18.9000	20.0750	0.362	18.0645	22.0855
16	40	20.0000	20.0750	0.362	18.0645	22.0855
17	35	.	18.7750	0.290	16.8187	20.7313

Obs	DENSITY	Residual
1	10	0.3250
2	10	-1.2750
3	10	-0.1750
4	10	-1.3750
5	20	0.4250
6	20	1.2250
7	20	0.0250
8	20	0.7250
9	30	0.4250
10	30	0.8250
11	30	1.1250
12	30	0.3250
13	40	-0.8750
14	40	-0.4750
15	40	-1.1750
16	40	-0.0750
17	35	.

Sum of Residuals		0
Sum of Squared Residuals		10.4700
Predicted Resid SS (Press)		14.0487

Figure 10.13. Graph of data with predicted values and prediction limits for plant density data

```
        -----+--------+--------+--------+--------+--------+--------+-----
   U95      |                                                       |
    22 +    |                                                  U    +
U        |                                                       |
p        |                                                       |
p        |                                                       |
e        |                                                       |
r        |                                                       |
         |                                                       |
B   20 + |                                                  ?    +
o        |                                                  1    |
u        |                                            U          |
n        |                                                  1    |
d        |                                                  1    |
         |                                            1          |
o        |                                            1          |
f   18 + |                                            1     L    +
         |                                            1          |
9        |                                            P          |
5        |                                                       |
%        |                              U                        |
         |                                                       |
C   16 + |                              1                        +
.        |                              1                        |
I        |                              1     L                  |
.        |                                                       |
(        |                              ?                        |
I        |                                                       |
n        |                                                       |
d        |    U                                                  |
i   14 + |                                                       +
v        |                                                       |
i        |                                                       |
d        |                              L                        |
u        |                                                       |
a        |    1                                                  |
l        |    P                                                  |
    12 + |    1                                                  +
P        |                                                       |
r        |                                                       |
e        |    1                                                  |
d        |    1                                                  |
)        |                                                       |
         |    L                                                  |
    10 + |                                                       +
        -----+--------+--------+--------+--------+--------+--------+-----
             10       15       20       25       30       35       40
```

DENSITY

10.4 Correlation

Example 10.19. To study the effects of air temperature on the growth characteristics of sour orange trees exposed to various concentrations of atmospheric carbon dioxide (CO_2), data were collected every other month, over a two-year period, on the dry weight per leaf and the mean air temperature of the preceding month, for trees exposed to an extra 300 μliter/liter of CO_2.

SAS's PLOT procedure can be used to obtain a graphical representation of the data, as shown in the program in Figure 10.14 and the graph in Figure 10.15. The graph indicates a relatively strong, positive association. Parameter estimates and the analysis of variance table resulting from fitting a straight line model relating leaf weight to temperature are in Figure 10.16. The ratio of the Model Sum of Squares

$$SS_{Y.X}[R] = 0.11426$$

to the Total Sum of Squares,

$$S_{Y.X}[TOT] = 0.18620$$

is shown in the analysis of variance portion of Figure 10.16 as

```
        R-square        0.6136,
```

indicating that a straight-line relationship of leaf weight to temperature can explain a little over 61% of the variation in leaf weight.

Figure 10.14. SAS program to compute the correlation coefficient

```
options ls=70 ps=54;
/*Correlation between air temp and dry leaf
      weight*/
data;
  input year @@;
    do i = 1 to 6;
      input temp grams @@;
      output;
    end;
  drop i;
lines;
90 12.0 1.00 18.4 1.10 23.2 1.16 26.4 1.28
      33.5 1.29 35.0 1.34
91 13.6 1.14 18.4 1.20 23.2 1.30 22.4 1.42
      33.5 1.39 33.6 1.38
;
run;
proc plot;
  plot grams*temp;
run;
proc reg;
  model grams=temp;
run;
proc corr;
  var grams temp;
```

Figure 10.17 shows output from the CORRelation procedure mentioned in the program in Figure 10.14. The CORRelation procedure first gives simple statistics for all variables mentioned in the VAR statement and then produces a matrix of bivariate correlation coefficients between all pairs of variables mentioned in the VAR statement. If the VAR statement is omitted, the procedure will compute statistics for every variable and correlations between all pairs of numerical variables in the data set.

Figure 10.15. Scatter plot of leaf weight versus temperature

```
       Plot of GRAMS*TEMP.   Legend: A = 1 obs,  B = 2 obs, etc.
GRAMS
 1.42 +                                          A
      |
 1.40 +
      |                                                          A
 1.38 +                                                          A
      |
 1.36 +
      |
 1.34 +                                                              A
      |
 1.32 +
      |
 1.30 +                                 A
      |                                                          A
 1.28 +                                      A
      |
 1.26 +
      |
 1.24 +
      |
 1.22 +
      |
 1.20 +                   A
      |
 1.18 +
      |
 1.16 +                              A
      |
 1.14 +          A
      |
 1.12 +
      |
 1.10 +                   A
      |
 1.08 +
      |
 1.06 +
      |
 1.04 +
      |
 1.02 +
      |
 1.00 +      A
      --+-----------+-----------+-----------+-----------+-----------+-
        10          15          20          25          30          35
                                  TEMP
```

Figure 10.16. Analysis of variance and parameter estimates for leaf dry weight data

```
Dependent Variable: GRAMS


                         Analysis of Variance

                        Sum of          Mean
Source           DF     Squares         Square      F Value      Prob>F

Model            1      0.11426        0.11426       15.881       0.0026
Error           10      0.07194        0.00719
C Total         11      0.18620

       Root MSE       0.08482     R-square        0.6136
       Dep Mean       1.25000     Adj R-sq        0.5750
       C.V.           6.78561


                        Parameter Estimates

                   Parameter        Standard      T for H0:
     Variable   DF  Estimate          Error      Parameter=0    Prob > |T|

     INTERCEP    1   0.941327      0.08123482       11.588        0.0001
     TEMP        1   0.012633      0.00317013        3.985        0.0026
```

Figure 10.17. Correlation matrix for leaf weight and temperature

```
                         Correlation Analysis

                 2 'VAR' Variables:   GRAMS      TEMP


                         Simple Statistics

     Variable        N      Mean    Std Dev      Sum    Minimum    Maximum

     GRAMS          12     1.2500    0.1301   15.0000    1.0000     1.4200
     TEMP           12    24.4333    8.0673    293.2    12.0000    35.0000


Pearson Correlation Coefficients / Prob > |R| under Ho: Rho=0 / N = 12


                              GRAMS                 TEMP

            GRAMS            1.00000              0.78334
                            0.0                   0.0026

            TEMP            0.78334              1.00000
                            0.0026                0.0
```

In this example, since there is only one pair of variables, a 2 × 2 matrix of correlation coefficients will be produced, as is shown in Figure 10.17. There are two rows to the matrix, one for each of the two variables, and there are also two columns. The order of the variables in the row and column headings is the same, and corresponds to the order in which the variables are listed in the VAR statement. (If no VAR statement had been used, the order of the variables would have corresponded to the order in the INPUT statement in the data step.)

There are two entries in each cell in the correlation matrix. The top entry is the correlation coefficient between the row variable and the column variable for that cell. Since any variable is perfectly and directly correlated with itself, all the correlations on the main diagonal (that set of cells for which the row number and column number are the same) of the matrix are +1.0000. Since the correlation of X with Y is the same as the correlation of Y with X, the correlation in the cell in the ith row and jth column is the same as the correlation in the jth row and the ith column; that is, the matrix is symmetric. In Figure 10.17 we see that the correlation coefficient is

$$r_{XY} = +0.78334,$$

which indicates a relatively strong positive association.

The bottom entry in each cell in the correlation matrix is a p-value for testing the significance of the correlation coefficient. The heading on the correlation matrix needs to be read carefully:

```
Pearson Correlation Coefficients / Prob > |R| under
               Ho:Rho = 0 / n = 12
```

This heading is taken to mean that the top entry in each cell is a Pearson product-moment correlation coefficient, and that under that is the p-value for testing the null hypothesis that the population correlation coefficient, ρ, is zero. The sample size, 12, is not shown in each cell, however, but only in this heading. These p-values are for the t-test as performed in Rao's Box 10.10:

$$t_{r_{XY}} = \frac{r_{XY}\sqrt{n-2}}{\sqrt{1-r_{XY}^2}}.$$

In Figure 10.17, the *p*-value, $p = 0.0026$, is the same as the *p*-value for both the *t*-test and the *F*-test of

$$H_0: \beta_1 = 0$$

performed in Figure 10.16, since for a simple linear model, all three hypotheses being tested are equivalent to

$$H_0: \text{no linear association.}$$

10.5 Determining Sample Sizes

Example 10.22. The performance of lungs is often measured by a quantity called the forced expiratory volume (FEV). There is known to be an association between FEV and height in humans. Suppose an investigator wants to verify the research hypothesis that the expected FEV for individuals who are 175 cm tall is more than 3 liters.

Let X and Y denote, respectively, the height and the FEV values for a human subject. Assume that the simple linear regression model

$$Y = \beta_0 + \beta_1 x + E,$$

where E is $N(0, \sigma^2)$, may be used to describe the relationship between FEV and height in healthy humans whose heights range between $a = 150$ cm and $b = 200$ cm. Then, with $\theta = \beta_0 + 175\beta_1$, the primary objective of the study is to test the null hypothesis $H_0: \theta = 3$ against the research hypothesis $H_1: \theta > 3$.

Suppose that the investigators want to use a 0.01-level t-test ($\alpha = 0.01$) and that they want at least 80% power if $\theta = 3.3$ or larger ($\Delta = 0.3$).

Suppose further that they expect the standard deviation σ of the error to be somewhere between 0.5 and 0.6.

In his Section 10.9, Rao provides SAS programs for computing the power of a one-sided t-test and the half-width of a 99% confidence interval on the parameter

$$\theta = c_0\beta_0 + c_1\beta_1 .$$

The noncentrality parameter, λ, as defined in Rao's Equation 10.18, is rather complex, as is the expression for the half-width, B in Equation 10.19. Consequently, you will probably want to copy and save the two programs and have them use data stored in a separate file.

Figure 10.18 shows the input data to perform the power analysis requested in Example 10.22. In the first two columns are the lower and upper limits, respectively, of the range of x-values of interest. In the third and fourth columns are the values for c_0 and c_1, respectively. Column 5 contains the value for Δ, column 6 holds the value for σ, column 7 is $t =$ the number of equally-spaced design points, and in the last column is n, the number of replications at each design point. In this example, we are investigating the power obtained by using 1, 2, 3 or 4 replications at 4 or 5 design points, when the standard deviation is either 0.5 or 0.6. These data are stored in a file called "nreg.dat."

Figure 10.19 shows the program for computing power for the one-sided 0.01-level test, and Figure 10.20 shows the output from this program. You might want to copy and test this program and then store it. To use this program in subsequent problems, all that is necessary is to create a new "nreg.dat" data file (and make appropriate changes if a different α-level or a two-sided test is required).

Figure 10.18. Data for regression power analysis

```
150 200 1 175 .3 .5 5 4
150 200 1 175 .3 .5 5 3
150 200 1 175 .3 .5 5 2
150 200 1 175 .3 .5 5 1

150 200 1 175 .3 .6 5 4
150 200 1 175 .3 .6 5 3
150 200 1 175 .3 .6 5 2
150 200 1 175 .3 .6 5 1

150 200 1 175 .3 .5 4 4
150 200 1 175 .3 .5 4 3
150 200 1 175 .3 .5 4 2
150 200 1 175 .3 .5 4 1

150 200 1 175 .3 .6 4 4
150 200 1 175 .3 .6 4 3
150 200 1 175 .3 .6 4 2
150 200 1 175 .3 .6 4 1
```

Figure 10.19. Program for computing power for a one-sided test

```
options ls=70 ps=54 pageno=1;
data a;
  infile 'nreg.dat';
  input a b c0 c1 delta sigma t n;
run;
data a;
  set a;
  nu     = n*t-2;
  m      = (delta**2)/(sigma**2);
  nr     = n*t*((b-a)**2)*(t+1);
  dr     = ((c0)**2)*((b-a)**2)*(t+1)+3*
           (t-1)*(2*c1-c0*(a+b))**2;
  lambda = (m*nr)/(2*dr);
  x      = tinv(0.99,nu);
  power  = 1-probt(x,nu,2*lambda);
run;
proc print;
  var n t nu sigma delta lambda power;
  title 'Power of One-Sided t-Test';
run;
```

Figure 10.20. Power analysis for a one-sided test

```
                    Power of One-Sided t-Test                          1

    OBS    N    T    NU    SIGMA    DELTA    LAMBDA    POWER

     1     4    5    18     0.5      0.3      3.600    0.99999
     2     3    5    13     0.5      0.3      2.700    0.99330
     3     2    5     8     0.5      0.3      1.800    0.74231
     4     1    5     3     0.5      0.3      0.900    0.11522
     5     4    5    18     0.6      0.3      2.500    0.98870
     6     3    5    13     0.6      0.3      1.875    0.84673
     7     2    5     8     0.6      0.3      1.250    0.40432
     8     1    5     3     0.6      0.3      0.625    0.06370
     9     4    4    14     0.5      0.3      2.880    0.99772
    10     3    4    10     0.5      0.3      2.160    0.91646
    11     2    4     6     0.5      0.3      1.440    0.46601
    12     1    4     2     0.5      0.3      0.720    0.05888
    13     4    4    14     0.6      0.3      2.000    0.89874
    14     3    4    10     0.6      0.3      1.500    0.60451
    15     2    4     6     0.6      0.3      1.000    0.22592
    16     1    4     2     0.6      0.3      0.500    0.03769
```

If $\sigma = 0.5$, it appears that the choices are to either use $n = 2$ replicates at each of $t = 5$ equally-spaced design points between 150 and 200 (giving power = 74.2%, a bit less than the desired 80%) or $n = 3$ replicates at each of $t = 4$ design points (power = 91.6%). If $\sigma = 0.6$, then $n = 3$ replicates at each of $t = 5$ design points or $n = 4$ replicates at each of $t = 4$ points will give power over 80%.

If interest centers on the precision with which an estimate can be made, rather than on testing an hypothesis about θ, then the program in Figure 10.21 can be used. You might want to copy and test this program and then save it. When the program in Figure 10.21 is applied to the data in Figure 10.18, the output in Figure 10.22 is produced.

Examination of the half-widths of the confidence intervals in Figure 10.22 gives you an idea of how closely you can expect to estimate the FEV of individuals 175 cm in height, depending on how much variability there is, and how many replications at how many design points you choose to use.

Figure 10.21. Program for power analysis for two-sided confidence intervals

```
options ls=70 ps=54 pageno=1;
data a;
  infile 'nreg.dat';
  input a b c0 c1 delta sigma t n;
run;
data a;
  set a;
  nu     = n*t-2;
  m      = (delta**2)/(sigma**2);
  nr     = n*t*((b-a)**2)*(t+1);
  dr     = ((c0)**2)*((b-a)**2)*(t+1)+3*
           (t-1)*(2*c1-c0*(a+b))**2;
  x      = tinv(0.995,nu);
  ratio  = dr/nr;
  b      = x*(ratio)**0.5*sigma;

run;
proc print;
  var n t nu sigma b;
  title 'Half-Widths of 99% Confidence
      Intervals';
run;
```

Figure 10.22. Power analysis for two-sided confidence intervals

Half-Widths of 99% Confidence Intervals

OBS	N	T	NU	SIGMA	B
1	4	5	18	0.5	0.32182
2	3	5	13	0.5	0.38888
3	2	5	8	0.5	0.53053
4	1	5	3	0.5	1.30607
5	4	5	18	0.6	0.38618
6	3	5	13	0.6	0.46666
7	2	5	8	0.6	0.63664
8	1	5	3	0.6	1.56728
9	4	4	14	0.5	0.37211
10	3	4	10	0.5	0.45745
11	2	4	6	0.5	0.65539
12	1	4	2	0.5	2.48121
13	4	4	14	0.6	0.44653
14	3	4	10	0.6	0.54893
15	2	4	6	0.6	0.78646
16	1	4	2	0.6	2.97745

10.6. Logistic Regression

Example 10.25. A suspected carcinogen, aflotoxin B_1, was fed to test animals. The following data for liver tumors were obtained...Let the response variable Y be defined as $Y = 1$ if the test animal developed a liver tumor and 0 otherwise. Then the tumor incidence data consist of $n = 18+22+22+21+25+28 = 130$ observations of the qualitative variable Y at $t= 6$ settings of the quantitative independent variable $X =$ dose. There are multiple responses at each setting of the independent variable...A logistic regression can be used to investigate the dose-response relationship between the probability of developing a liver tumor and the dose of the carcinogen.

The LOGISTIC regression procedure in SAS works very much like the REGression procedure, in terms of specifying the model, the types of commands used, and the options available. The LOGISTIC procedure can be used to analyze data for either a binary response variable, as is the case in this example, or for a nominal response variable with more than two ordered categories (for example, "agree, neutral, disagree"). Also, data may include multiple responses measured at each value of the independent variable, as is the case in this example, or the data may consist of only one response at several levels of the independent variable. How to analyze data of the latter type will be illustrated using the data of Example 10.24.

Figure 10.23 shows a program for analyzing the tumor incidence data. As long as the response variable is binary, there are two choices for how to read in the data. In Figure 10.23, a case is a single dosage level, and the information given for each case is what the dose was, how many animals received that dose, and how many of those animals developed tumors. Then the MODEL statement in the LOGISTIC procedure states that the *fraction* of animals

```
tumors/animals
```

Figure 10.23. Logistic regression program for tumor incidence data

```
options ls=70 ps=54;
data dose;
  input dose animals tumors;
  lines;
    0 18  0
    1 22  2
    5 22  1
   15 21  4
   50 25 20
  100 28 28
  ;
run;
proc logistic data=dose;
  model tumors/animals=dose;
  output out=est p=pred l=lower u=upper;
run;
proc print data=est;
run;
```

which developed tumors in any group is a function of the dose administered to animals in that group. Another way to read in data will be illustrated later with Example 10.24. Figure 10.24 shows the logistic regression analysis. Portions of this output are extracted in the figure in Rao's Example 10.26.

At the top of Figure 10.24 we see some summary information: 55 animals developed tumors and 81 did not. Under "Criteria for Assessing Model Fit," the first two statistics, AIC and SC, are *the Akaike Information Criterion* and the *Schwartz Criterion*, respectively. These statistics are used for comparing different models fitted to the same data (a lower value indicates a "better" model). The *Score* statistic and *the – 2 LOG L* statistic both provide large sample tests of

H_0: the model explains no variation in the response.

Generally, these two statistics, while differing somewhat, will give the same conclusion to the test. In this case, with either test statistic, we would reject the null hypothesis since $p = 0.0001$ for both and conclude that the probability of tumor occurrence is a function of dosage level.

Figure 10.24. Logistic regression analysis for tumor incidence data

```
The LOGISTIC Procedure

Data Set: WORK.DOSE
Response Variable (Events): TUMORS
Response Variable (Trials): ANIMALS
Number of Observations: 6
Link Function: Logit

                        Response Profile

                   Ordered  Binary
                   Value    Outcome      Count

                     1      EVENT          55
                     2      NO EVENT       81

               Criteria for Assessing Model Fit

                                 Intercept
                     Intercept      and
   Criterion          Only       Covariates   Chi-Square for Covariates

   AIC               185.535       73.907          .
   SC                188.447       79.733          .
   -2 LOG L          183.535       69.907      113.627 with 1 DF (p=0.0001)
   Score               .             .           87.493 with 1 DF (p=0.0001)

              Analysis of Maximum Likelihood Estimates

                 Parameter  Standard     Wald      Pr >    Standardized
     Variable DF  Estimate    Error   Chi-Square Chi-Square   Estimate

     INTERCPT 1    -3.0360   0.4823    39.6336    0.0001         .
     DOSE     1     0.0901   0.0146    38.3089    0.0001     1.907429

        Analysis of
          Maximum
        Likelihood
         Estimates


                    Odds
     Variable       Ratio

     INTERCPT       0.048
     DOSE           1.094

The LOGISTIC Procedure
    Association of Predicted Probabilities and Observed Responses

                   Concordant = 92.0%     Somers' D = 0.891
                   Discordant =  2.9%     Gamma     = 0.939
                   Tied       =  5.1%     Tau-a     = 0.432
                   (4455 pairs)           c         = 0.946
```

Under "Analysis of Maximum Likelihood Estimates" we find estimates of the parameters of the logistic regression equation, so that the probability of occurrence of a tumor for dosage x of the carcinogen is given by

$$\hat{P}(x) = \frac{e^{-3.0360+0.0901x}}{1 + e^{-3.0360+0.0901x}} \ .$$

These probabilities are shown in Figure 10.25 at each of the dosage levels in the experiment. (How Figure 10.25 was produced will be explained shortly). Note in Figure 10.24 that a different method for obtaining parameter estimates is used in LOGISTIC than is used in REG: these are "maximum likelihood", rather than "least squares" estimates; hence the necessity for a different SAS procedure to analyze categorical response variables.

Standard errors of the estimates are given, and then Wald Chi-Square statistics, which test whether the estimates differ significantly from zero. To compute the value of the Wald statistic for either estimate, divide the parameter estimate by its standard error; then square the result, so that this statistic is like the square of a t-statistic. Both p-values being 0.0001 in Figure 10.24, we conclude that both the "intercept" and the "slope" estimates do differ significantly from zero. Computation of the "standardized estimate" of the "slope" parameter, 1.907429, is not obvious from other quantities of the printout. Standardized estimates are of value in assessing the relative effects of multiple independent variables on the response, when the different independent variables have different units of measurement.

Figure 10.25. Predicted probabilities of tumor occurrence as a function of dose, with upper and lower 95% confidence limits

OBS	DOSE	ANIMALS	TUMORS	PRED	LOWER	UPPER
1	0	18	0	0.04582	0.01832	0.10999
2	1	22	2	0.04993	0.02043	0.11691
3	5	22	1	0.07007	0.03141	0.14900
4	15	21	4	0.15647	0.08559	0.26880
5	50	25	20	0.81281	0.62719	0.91809
6	100	28	28	0.99746	0.97686	0.99973

The values for Odds Ratios for the intercept and DOSE, respectively, under "Analysis of Maximum Likelihood Estimates" are

$$0.048 = e^{-3.036}$$

and

$$1.094 = e^{0.0901} .$$

The odds in favor of an event is defined as the probability that the event will occur divided by the probability that the event will not occur. That is, for dosage $x = 0$, the odds that a tumor will occur are 0.048, or equivalently, the odds against a tumor are the reciprocal, $1/0.048 = 20.8$. Thus, at dosage $x = 0$, it is almost 21 times more likely that a tumor will not occur than that it will. With each increase in dosage of 1 ppb, the odds of occurrence of a tumor increase by a factor of 1.094.

Finally, in Figure 10.24, we see an analysis of the association between the predicted probabilities of tumor occurrence and the actual occurrences, that is, an analysis of how well the logistic regression model fits the data. These measures are actually more appropriate for ordinal, rather than binary response. For each case in the data, SAS computes the probability of tumor occurrence and on the basis of this probability, classifies each case as to whether or not a tumor would be expected. Then, for every pair of cases in the data, SAS computes whether the pair is concordant or discordant. A pair of cases is concordant if one of the pair is higher than the other on the predicted probability of occurrence, then that member of the pair is also "higher" than the other on whether or not a tumor was observed ("higher" meaning yes). A pair of cases is discordant if the member with the higher predicted probability does not develop a tumor, while the one with the lower probability does. The percentage of concordant or discordant pairs may be interpreted like a coefficient of determination, in that the closer to 100%, the greater the predictive ability of the model. If there is a large percentage of concordant pairs, then the likelihood of occurrence of the event increases with X; a preponderance of discordant pairs indicates that the probability of occurrence decreases as the value of X

increases. Somers' *D*, Gamma, and *c* are similar kinds of correlation measures. Tau-a also measures degree of association, but on a different scale.

Let us return to Figure 10.25 and consider the code that was used to produce it. In order to produce the printout in Figure 10.24, SAS had to compute the predicted probabilities for each case in the sample. These predicted probabilities, along with upper and lower confidence limits, can be added to the data set that was used to perform the regression, by using the OUTPUT command.

To use the OUTPUT statement, you must specify what the name of the newly-created data set is going to be and the names you will use to refer to the variables. SAS refers to the output data set as OUT=, the predicted probabilities as P=, and upper and lower confidence limits as U= and L=, respectively. In Figure 10.23, the user-supplied name for the output data set is called EST (for estimates), predicted values are called PRED, and upper and lower limits are called UPPER and LOWER, respectively. A look at the LOG file for the program in Figure 10.23 gives this message after the DATA step:

```
The data set WORK.DOSE has 6 observations and 3
                    variables.
```

The three variables are DOSE, ANIMALS, and TUMORS. In the LOG file after the REG procedure is the message:

```
The data set WORK.EST has 6 observations and 6 variables.
```

The six variables in the data set WORK.EST are the three in the WORK.DOSE data set, plus PRED, UPPER, and LOWER.

Now that the predicted values and upper and lower confidence limits have been included in a data set, they can be used in other procedures, namely the PRINT and PLOT procedures. The next-to-last section of the program in Figure 10.23 produced the listing of predicted probabilities in Figure 10.25, and the last section of the program in Figure 10.23 produced the graph of the logistic regression function, with upper and lower confidence bands, which is shown in Figure 10.26. Connecting the asterisks in Figure 10.26 produces an *S*-shaped curve, showing that the probability of incidence of tumors increases sharply when the dose level reaches about 20 ppb.

Figure 10.26. Plot of logistic regression function for tumor incidence

```
                    Plot of PRED*DOSE.    Symbol used is '*'.
                    Plot of LOWER*DOSE.   Symbol used is 'L'.
                    Plot of UPPER*DOSE.   Symbol used is 'U'.

            |
            |
     1.0  + |                                                        *
            |                                                        L
            |
            |                                 U
   E        |
   s        |
   t  0.8 + |                                 *
   i        |
   m        |
   a        |
   t        |
   e        |                                 L
   d  0.6 + |
            |
   P        |
   r        |
   o        |
   b        |
   a  0.4 + |
   b        |
   i        |
   l        |
   i        |             U
   t        |
   y  0.2 + |
            |          *
            |     U  U
            |   U         L
            |       *
            |   ** L
     0.0  +
            |
          --+----------+----------+----------+----------+----------+-
            0         20         40         60         80        100

                                    DOSE

   NOTE: 3 obs hidden.
```

Example 10.24. Bone marrow transplantation (BMT) is a curative therapy for acute leukemias and other blood-related diseases. One problem associated with BMT is that medications administered when the patient is being prepared for marrow transplantation can cause life threatening toxicities. Let X be the age (in years) of the BMT patient, and Y be a binary variable defined as $Y = 1$ if the patient has acute pulmonary toxicity within 100 days after BMT, and $Y = 0$ if otherwise.

The survival times and events associated with various drug toxicities for 73 BMT patients suffering from acute lymphocytic leukemia (ALL) were analyzed...for 24 of the patients.

In the previous example, there were only six different values of the independent variable, and multiple cases at each value. In this example, there are many different values of the independent variable, age, and only one patient per age in many instances. In the program in Figure 10.27, a case is a patient, and there are two pieces of information for each patient: age and whether or not pulmonary toxicity developed. The binary response variable, TOXIC, was entered as a character variable, but it could have been coded as yes = 1 and no = 0 or with any other convenient coding.

Recall that although the REGression procedure had "built-in" commands for producing printouts and plots, the OUTPUT command could have been used in that procedure, as well.

The MODEL statement in this figure is very much like the MODEL statement in the REG procedure, and the rest of the program in Figure 10.27 is very similar to the one in Figure 10.23. The data were sorted by AGE so that the case-by-case printout of the predicted values would be done in an orderly fashion; the data were then sorted by TOXIC so that the value "no" was the first one in the data set, since character values will be arranged in alphabetical order. There has to be some way to tell SAS that the value "yes" means that the event *did* occur. This is done by the ORDER statement in the LOGISTIC command: ORDER=DATA, meaning that the data have been read in such a way that the lowest value of the response variable is entered first, followed by the next lowest (and so on, if the response variable were ordinal instead of binary). Note carefully (this is tricky!) that SAS uses the *last* level of the response variable as the reference. Always refer to the LOG file to be sure you know what value is being considered occurrence of the event. A portion of the LOG file corresponding to the program in Figure 10.27 is shown in Figure 10.28. If you had received the message that LOGISTIC is modeling the probability that TOXIC = "no", then simply include the command DESCENDING in the PROC statement after the ORDER=DATA.

Figure 10.27. Logistic regression program to analyze toxicity data

```
options ls=70 ps=54;
data a;
  input age toxic$ @@;
lines;
34 no 13 yes 31 yes 10 yes 13 no 29 no 7 yes
16 yes 25 yes 26 no 14 yes 55 no 4 no 46 no
24 no 15 no 32 yes 10 yes 3 yes 5 yes 5 yes
21 no 35 yes 5 no
;
run;
proc sort data=a;
  by age toxic;
run;
proc logistic data=a order=data;
  model toxic=age;
  output out=est p=pred l=lower u=upper;
run;
proc print data=est;
run;
proc plot data=est;
  plot pred*age='*' lower*age='L'
       upper*age='U'/overlay;
run;
```

Figure 10.28. Portion of SAS LOG file for logistic regression

```
NOTE: PROC LOGISTIC is modeling the probability that TOXIC='yes'. One
      way to change this to model the probability that TOXIC='no' is
      to specify the DESCENDING option on the PROC statement. Refer
      to Technical Report P-229 or the SAS System Help Files for
      details.
```

Figure 10.29. Logistic regression analysis of toxicity as a function of age

```
Data Set: WORK.A
Response Variable: TOXIC
Response Levels: 2
Number of Observations: 24
Link Function: Logit

                          Response Profile

                   Ordered
                     Value   TOXIC        Count

                        1    yes            13
                        2    no             11


                   Criteria for Assessing Model Fit

                              Intercept
                   Intercept     and
Criterion           Only      Covariates   Chi-Square for Covariates

AIC                35.104      34.504          .
SC                 36.282      36.860          .
-2 LOG L           33.104      30.504       2.600 with 1 DF  (p=0.1068)
Score                .            .          2.501 with 1 DF  (p=0.1138)


               Analysis of Maximum Likelihood Estimates

           Parameter Standard    Wald        Pr >     Standardized
  Variable DF Estimate  Error  Chi-Square Chi-Square    Estimate

  INTERCPT 1    1.1938  0.8007   2.2227     0.1360           .
  AGE      1   -0.0515  0.0343   2.2602     0.1327     -0.397698

      Analysis of
        Maximum
      Likelihood
       Estimates


                Odds
  Variable     Ratio

  INTERCPT     3.300
  AGE          0.950

Association of Predicted Probabilities and Observed Responses

                Concordant = 64.3%       Somers' D = 0.308
                Discordant = 33.6%       Gamma     = 0.314
                Tied       =  2.1%       Tau-a     = 0.159
                (143 pairs)              c         = 0.654
```

The logistic regression analysis of the toxicity data is shown in Figure 10.29. The *p*-values slightly larger than 0.10 indicate not very strong evidence that incidence of toxicity varies with the age of the patient. Nevertheless, predicted probabilities of incidence are given in Figure 10.30 and are plotted in Figure 10.31. Note how wide apart the confidence bands are in Figure 10.31. The column of "yesses" labeled

LEVEL

in Figure 10.30 indicates that the probabilities in the next column are the predicted probabilities of the case belonging to the "yes" category.

Figure 10.30. Predicted probabilities of toxicity as a function of age, with 95% confidence limits

OBS	AGE	TOXIC	_LEVEL_	PRED	LOWER	UPPER
1	3	yes	yes	0.73870	0.40977	0.92008
2	4	no	yes	0.72864	0.41011	0.91205
3	5	no	yes	0.71833	0.41012	0.90343
4	5	yes	yes	0.71833	0.41012	0.90343
5	5	yes	yes	0.71833	0.41012	0.90343
6	7	yes	yes	0.69703	0.40903	0.88435
7	10	yes	yes	0.66343	0.40386	0.85154
8	10	yes	yes	0.66343	0.40386	0.85154
9	13	no	yes	0.62811	0.39292	0.81506
10	13	yes	yes	0.62811	0.39292	0.81506
11	14	yes	yes	0.61599	0.38766	0.80255
12	15	no	yes	0.60374	0.38147	0.79008
13	16	yes	yes	0.59135	0.37430	0.77781
14	21	no	yes	0.52797	0.32272	0.72418
15	24	no	yes	0.48936	0.28095	0.70154
16	25	yes	yes	0.47650	0.26594	0.69575
17	26	no	yes	0.46367	0.25068	0.69079
18	29	no	yes	0.42553	0.20499	0.68029
19	31	yes	yes	0.40055	0.17604	0.67635
20	32	yes	yes	0.38825	0.16235	0.67513
21	34	no	yes	0.36408	0.13690	0.67390
22	35	yes	yes	0.35224	0.12522	0.67382
23	46	no	yes	0.23580	0.04157	0.68700
24	55	no	yes	0.16253	0.01537	0.70705

Figure 10.31. Plot of logistic regression function and confidence bands for the probability of toxicity as a function of age

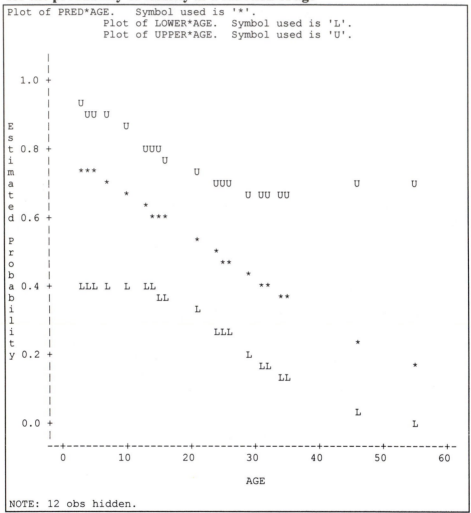

```
Plot of PRED*AGE.    Symbol used is '*'.
              Plot of LOWER*AGE.   Symbol used is 'L'.
              Plot of UPPER*AGE.   Symbol used is 'U'.

         |
         |
     1.0 +
         |
         |   U
         |     UU U
E        |           U
s        |
t   0.8  +           UUU
i        |             U
m        |   ***
a        |      *         U
t        |        *     UUU
e        |             U UU UU        U        U
d   0.6  +       *
         |        ***
P        |
r        |             *
o        |              *
b        |               **
a   0.4  +   LLL L   L   LL      *
b        |             LL        **
i        |                 L        **
l        |
i        |             LLL
t        |
y   0.2  +           L
         |            LL
         |             LL
         |
         |
         |                      *
     0.0 +                        L
         |
        --+---------+---------+---------+---------+---------+---------+-
          0        10        20        30        40        50        60

                                AGE
NOTE: 12 obs hidden.
```

11

Multiple Linear Regression Analysis

11.1 Introduction

Using the REG or the GLM procedure in SAS, the analysis of a multiple linear regression model is hardly appreciably more difficult than the analysis of a simple linear regression model. The GLM procedure allows for a sequential analysis of the variables in a multiple regression model.

11.2 Estimation of Parameters

Example 11.1. Enzymes are organic substances produced by living cells. Because enzymes are capable of inducing chemical changes in other substances without themselves being changed in the process, biologists are very interested in the capacity of enzymes to adhere to substances found in the human body.

In an experiment on the adsorption (adhesion to a solid) of a particular enzyme on ultrafine particles of silica suspended in a liquid medium, two variables were measured: C, the concentration of the enzyme (g/dm^3) in the medium; and Y, the amount of the enzyme adsorbed (expressed as a percentage of the amount of silica in the medium).

A primary objective of the study was to see how the amount of the enzyme adsorbed varied with its concentration in the medium.

Figure 11.1. **Amount of enzyme adsorbed as a function of enzyme concentration in the medium**

```
              Plot of Y*C.   Legend: A = 1 obs, B = 2 obs, etc.

    Y |
      |
   25 +
      |
      |                                                                A
      |
      |                                                       A
      |
      |
   20 +                                            A
      |
      |                                     A
      |
      |
      |
      |
   15 +
      |                          A
      |
      |
      |
      |
   10 +
      |
      |
      |
      |
      |
    5 +            A
      |
      |
      |
      |     A
      |
      |
    0 +
      |
    ---+-----+-----+-----+-----+-----+-----+-----+-----+-----+-----+--
      0.05  0.15  0.25  0.35  0.45  0.55  0.65  0.75  0.85  0.95  1.05
                                       C
```

A plot of the data, as in Figure 11.1, indicates that the relationship of the amount the enzyme adsorbed to the concentration of the enzyme in the medium is a curvilinear one. (Refer to the program in Figure 11.2 for the code that produced Figure 11.1.) Accordingly, a second-degree polynomial, or quadratic, function is used to describe the relationship. As in Rao's Example 11.5, the model of the relationship is thus

$$Y_i = \beta_0 + \beta_1 x_{i1} + \beta_2 x_{i2} + E_i$$

where

Y_i = the adsorbed amount of enzyme (%)

x_{i1} = the concentration of the enzyme in the i-th medium

$x_{i2} = x_{i1}^2,$

and

E_i = the random error in the i-th measured response.

Figure 11.2. SAS program to fit a quadratic equation to the enzyme adsorption data

```
options ls=70 ps=54;
/*Quadratic Regression*/
data;
   input c y @@;
   csq=c**2;
lines;
.05 2.0 .10 5.0 .30 14.5 .50 18.6 .60 20.1
.80 22.0 1.00 23.5
;
run;
proc plot;
   plot y*c;
run;
proc reg;
   model y = c csq/p;
   plot y*c p.*c='*'/overlay;
run;
```

There is one *explanatory* variable in the above model, namely the concentration of the enzyme in the medium. Explanatory variables are those listed, along with the response variable, the INPUT statement of a SAS program. There are two *independent* variables, namely the concentration and the square of the concentration. Independent variables are those listed in the MODEL statement of the REG or the GLM procedure, as is shown in Figure 11.2. In order to be able to use the independent variables in a MODEL statement, however, they must be contained in the data set being used in the regression analysis. Thus, in the DATA step of the program in Figure 11.2, the independent variable "CSQ" is defined by squaring the values of the explanatory variable, "*C* ". A printout of the data set would show values for Y_i, x_{i1}, and $x_{i2} = x_{i1}^2$ as in the table in Rao's Example 11.5.

The output from the regression analysis is shown in Figure 11.3. This output corresponds to that presented in Rao's Example 11.9 and the analysis of variance table in his Example 11.11. The estimated regression equation is thus

$$\hat{y} = 0.17 + 51.25x - 28.58x^2.$$

Predicted values are plotted with the data in Figure 11.4.

Figure 11.3. **Regression analysis for the quadratic model (REG procedure)**

```
                          Analysis of Variance

                       Sum of          Mean
Source          DF     Squares         Square      F Value      Prob>F

Model            2    425.13863      212.56931     199.531      0.0001
Error            4      4.26137        1.06534
C Total          6    429.40000

        Root MSE        1.03215      R-square      0.9901
        Dep Mean       15.10000      Adj R-sq      0.9851
        C.V.            6.83546

                          Parameter Estimates

                    Parameter      Standard     T for H0:
Variable    DF       Estimate         Error     Parameter=0    Prob > |T|

INTERCEP     1       0.172006     0.93348130        0.184        0.8628
C            1      51.261041     4.48292181       11.435        0.0003
CSQ          1     -28.577485     4.28710269       -6.666        0.0026


                      Dep Var    Predict
                 Obs     Y        Value     Residual

                   1   2.0000     2.6636    -0.6636
                   2   5.0000     5.0123    -0.0123
                   3  14.5000    12.9783     1.5217
                   4  18.6000    18.6582    -0.0582
                   5  20.1000    20.6407    -0.5407
                   6  22.0000    22.8912    -0.8912
                   7  23.5000    22.8556     0.6444

Sum of Residuals                       0
Sum of Squared Residuals          4.2614
Predicted Resid SS (Press)       24.0925
```

Figure 11.4. Quadratic equation plotted with data for enzyme adsorption example

```
    ---+-----+-----+-----+-----+-----+-----+-----+-----+-----+-----+---
  Y |                                                                 |
    |                                                                 |
    |                                                                 |
    |                                                                 |
 25 +                                                                 +
    |                                                                 |
    |                                                              1  |
    |                                                  *           *  |
    |                                                  1              |
    |                                                                 |
    |                                      *                          |
 20 +                                      1                          +
    |                                                                 |
    |                                ?                                |
    |                                                                 |
    |                                                                 |
    |                                                                 |
    |                                                                 |
 15 +                                                                 +
    |                    1                                            |
    |                                                                 |
    |                    *                                            |
    |                                                                 |
    |                                                                 |
    |                                                                 |
 10 +                                                                 +
    |                                                                 |
    |                                                                 |
    |                                                                 |
    |                                                                 |
    |                                                                 |
    |                                                                 |
  5 +        ?                                                        +
    |                                                                 |
    |                                                                 |
    |     *                                                           |
    |     1                                                           |
    |                                                                 |
    |                                                                 |
  0 +                                                                 +
    |                                                                 |
    |                                                                 |
    |                                                                 |
    |                                                                 |
    ---+-----+-----+-----+-----+-----+-----+-----+-----+-----+-----+---
      0.0   0.1   0.2   0.3   0.4   0.5   0.6   0.7   0.8   0.9   1.0

                                     C
```

11.3 Inferences about Expected and Future Responses

Example 11.5. In Example 11.2 we described a study on oxygen demand in dairy wastes. In this study the response variable is $Y = \log$ (oxygen demand) (the oxygen demand is measured in mg/liter). The five explanatory variables are: X_1, the biological oxygen demand (g/liter); X_2, the total Kjeldahl nitrogen (g/liter); X_3, the total solids (g/liter); X_4, the total volatile solids (g/liter); and X_5, the chemical oxygen demand (g/liter).

For purposes of illustration, we'll use a multiple linear regression analysis to model the biological oxygen demand as a function of X_1 and X_2 only.

This example differs from the previous one in that neither of the two independent variables, X_1 and X_2, is a mathematical function of the other. That is, both of the independent variables are explanatory variables; both are listed in the INPUT statement in the data step in the program in Figure 11.5. The REG procedure can handle with equal ease independent variables which are separate explanatory variables and/or functions of explanatory variables.

Figure 11.5. Multiple regression program with confidence and prediction limits

```
options ls=70 ps=54;
/*Multiple Regression*/
data;
   infile 'manure.dat';
   input day bioox nitrogen tsolids vsolids
chemox demand;
run;
proc reg;
   model demand=bioox nitrogen/cli clm;
   id day;
run;
```

Similarly, predicted values, residuals, and confidence or prediction limits for a multiple regression model can be obtained from the REG procedure in exactly the same way as for a simple linear model (see the program in Figure 10.10). However, note that since there are two explanatory variables and one response variable, this multiple regression model is a *three-dimensional* one. Two-dimensional plots, as were made in Chapter 10 and in Example 11.1, are no longer appropriate. Figures 11.6 and 11.7 show the output from the program in Figure 11.5.

Figure 11.6. Analysis of variance and parameter estimates for multiple regression

```
                        Analysis of Variance

                        Sum of          Mean
Source          DF      Squares         Square      F Value      Prob>F

Model            2      3.48697        1.74349       18.748       0.0001
Error           17      1.58094        0.09300
C Total         19      5.06792

        Root MSE        0.30495      R-square      0.6880
        Dep Mean        0.11920      Adj R-sq      0.6513
        C.V.          255.83361

                        Parameter Estimates

                      Parameter      Standard     T for H0:
    Variable   DF      Estimate         Error    Parameter=0     Prob > |T|

    INTERCEP   1      -1.420604      0.36669193      -3.874        0.0012
    BIOOX      1       1.422058      0.23361056       6.087        0.0001
    NITROGEN   1       2.907873      1.33480559       2.178        0.0437
```

Figure 11.7. Predicted values, confidence and prediction limits

Obs	DAY	Dep Var DEMAND	Predict Value	Std Err Predict	Lower95% Mean	Upper95% Mean
1	0	1.5563	0.8538	0.138	0.5626	1.1451
2	7	0.8976	0.6670	0.124	0.4047	0.9293
3	15	0.7482	0.5548	0.115	0.3126	0.7971
4	22	0.7160	0.6906	0.116	0.4454	0.9358
5	29	0.3130	0.7731	0.135	0.4875	1.0586
6	37	0.3617	0.5746	0.106	0.3519	0.7974
7	44	0.1139	0.3090	0.090	0.1192	0.4987
8	58	0.1139	0.0853	0.073	-0.0684	0.2390
9	65	-0.2218	0.0129	0.086	-0.1688	0.1947
10	72	-0.1549	-0.1117	0.103	-0.3285	0.1050
11	80	0	-0.1709	0.118	-0.4203	0.0784
12	86	0	-0.1128	0.098	-0.3193	0.0938
13	93	-0.0969	0.0113	0.077	-0.1521	0.1746
14	100	-0.2218	-0.2039	0.124	-0.4649	0.0571
15	107	-0.3979	0.0652	0.111	-0.1689	0.2993
16	122	-0.1549	-0.4538	0.117	-0.7010	-0.2066
17	129	-0.2218	-0.1254	0.083	-0.3012	0.0504
18	151	-0.3979	-0.5753	0.132	-0.8545	-0.2961
19	171	-0.5229	-0.1227	0.164	-0.4686	0.2231
20	220	-0.0458	-0.3370	0.186	-0.7302	0.0562

Obs	DAY	Lower95% Predict	Upper95% Predict	Residual
1	0	0.1476	1.5601	0.7025
2	7	-0.0278	1.3618	0.2306
3	15	-0.1326	1.2423	0.1934
4	22	0.00207	1.3792	0.0254
5	29	0.0691	1.4770	-0.4601
6	37	-0.1062	1.2555	-0.2129
7	44	-0.3618	0.9798	-0.1951
8	58	-0.5762	0.7468	0.0286
9	65	-0.6556	0.6815	-0.2347
10	72	-0.7907	0.5672	-0.0432
11	80	-0.8610	0.5191	0.1709
12	86	-0.7885	0.5629	0.1128
13	93	-0.6526	0.6751	-0.1082
14	100	-0.8982	0.4904	-0.0179
15	107	-0.6195	0.7499	-0.4631
16	122	-1.1430	0.2355	0.2989
17	129	-0.7924	0.5416	-0.0964
18	151	-1.2767	0.1261	0.1774
19	171	-0.8532	0.6077	-0.4002
20	220	-1.0911	0.4170	0.2912

Sum of Residuals	0
Sum of Squared Residuals	1.5809
Predicted Resid SS (Press)	2.4787

11.4 Testing Simultaneous Hypotheses

Example 11.18. [Rao's] Table 11.4 presents data on the percentages of nitrogen (X_1), chlorine (X_2), potassium (X_3), and the log of leafburn time in seconds (Y) for a sample of $n = 30$ tobacco leaves.

Assume the multiple linear regression model

$$Y = \beta_0 + \beta_1 x_1 + \beta_2 x_2 + \beta_3 x_3 + E$$

and suppose that we want to test the null hypothesis that addition of chlorine and potassium will not increase the predictive power of a model that contains only nitrogen as the independent variable—in symbols, the hypothesis $H_0: \beta_2 = \beta_3 = 0$. The reduced model under H_0 is

$$Y = \beta_0 + \beta_1 x_1 + E .$$

Example 11.18 in Rao shows the analysis from the REG procedure for both the three-predictor and the one-predictor models described in Example 11.18. The reduction in the error sum of squares by adding X_2 and X_3 to the model containing X_1 can be computed using SS[E] from the two analyses. Rather than computing the reduction in error sum of squares and the associated F-test by hand, however, it is possible to have SAS test the null hypothesis that neither X_2 nor X_3 improves over X_1 alone. The program in Figure 11.8 shows how the TEST command is used to perform this test.

The TEST command in the REG procedure can be used to test a wide array of null hypotheses about the regression parameters. The general form of the command is

```
test varname1=c1, varname2=c2, ..., varnamek=ck;
```

Figure 11.8. Using the TEST statement in the REG procedure

```
options ls=70 ps=54;
/*Reduced Model*/
data;
   infile 'torrie.dat';
   input nitrogen chlorine potas logtime;
run;
proc reg;
   model logtime=nitrogen chlorine potas;
   A: test chlorine, potas;
run;
```

Figure 11.9. Regression analysis and test of full vs. reduced model

```
                    Analysis of Variance

                        Sum of        Mean
Source          DF     Squares       Square     F Value     Prob>F

Model            3     5.50473      1.83491      40.267      0.0001
Error           26     1.18479      0.04557
C Total         29     6.68952

       Root MSE       0.21347    R-square      0.8229
       Dep Mean       0.68600    Adj R-sq      0.8025
       C.V.          31.11782


                    Parameter Estimates

                 Parameter      Standard    T for H0:
Variable   DF     Estimate         Error   Parameter=0    Prob > |T|

INTERCEP    1     1.811043    0.27951935        6.479        0.0001
NITROGEN    1    -0.531455    0.06957678       -7.638        0.0001
CHLORINE    1    -0.439636    0.07303727       -6.019        0.0001
POTAS       1     0.208975    0.04064022        5.142        0.0001


Dependent Variable: LOGTIME
Test: A         Numerator:     1.0294  DF:    2   F value: 22.5893
                Denominator:  0.045569  DF:   26   Prob>F:   0.0001
```

where c_1, c_2, ..., c_k are constants. If c_1, c_2, ..., c_k are zero, they may be omitted, as is shown in the program in Figure 11.8. Several TEST statements can be used in a single regression procedure. To keep which output goes with which test command straight in such instances, a provision is made for labeling the output by writing the name of the label followed by a colon. In the example in Figure 11.8, the test is labeled simply as "A." Up to five characters can be used in a label.

Figure 11.9 shows the regression analysis for the full (three-predictor) model and the test of whether the coefficients on both chlorine and potassium differ significantly from zero. Compare the F-value to the one obtained in Rao's Example 11.18.

11.5 Multiple and Partial Correlation Coefficients

Example 11.23. Suppose that we want to fit the model

$$Y = \beta_0 + \beta_1 x_1 + \beta_2 x_2 + \beta_3 x_3 + E$$

in Example 11.18 to the data on Y = log (leafburn time), X_1 = percentage of nitrogen, X_2 = percentage of chlorine and X_3 = percentage of potassium.

Either the REG or the GLM procedure can be used to perform multiple linear regression in SAS. Much of the output is the same in both procedures, but the GLM procedure gives, in addition to the analysis of variance table and parameter estimates, partial ("Type III") and sequential ("Type I") sums of squares and F-tests. The code

```
proc glm;
model logtime=nitrogen chlorine potas;
```

produces the GLM analysis of the Steel and Torrie leafburn data in Figure 11.10. Portions of Figure 11.10 are shown in Rao's Example 11.23.

Figure 11.10. GLM analysis of leafburn data

```
                      General Linear Models Procedure

Dependent Variable: LOGTIME

Source                    DF    Sum of Squares    F Value       Pr > F

Model                      3        5.50473408      40.27       0.0001

Error                     26        1.18478592

Corrected Total           29        6.68952000

                  R-Square              C.V.            LOGTIME Mean

                  0.822889           31.11782            0.68600000


Source                    DF        Type I SS    F Value       Pr > F

NITROGEN                   1        3.44600764      75.62       0.0001
CHLORINE                   1        0.85384481      18.74       0.0002
POTAS                      1        1.20488163      26.44       0.0001

Source                    DF      Type III SS    F Value       Pr > F

NITROGEN                   1        2.65871240      58.35       0.0001
CHLORINE                   1        1.65106256      36.23       0.0001
POTAS                      1        1.20488163      26.44       0.0001


                                  T for H0:    Pr > |T|    Std Error of
Parameter         Estimate     Parameter=0                   Estimate

INTERCEPT       1.811042531           6.48      0.0001       0.27951935
NITROGEN       -0.531455299          -7.64      0.0001       0.06957678
CHLORINE       -0.439635791          -6.02      0.0001       0.07303727
POTAS           0.208975307           5.14      0.0001       0.04064022
```

11.6 Selecting Independent Variables

Example 11.24. For the leafburn data in Example 11.18 we want to select a set of independent variables to model $Y = \log$ (leafburn time) as a function of three variables: X_1, the percentage of nitrogen; X_2, the percentage of chlorine; and X_3, the percentage of potassium. For the purpose of this example, let's assume that $\{X_1, X_2, X_3, X_1^2, X_2^2, X_3^2, X_1 X_2, X_1 X_3, X_2 X_3\}$ is the candidate set of independent variables. Our objective is to use the R^2 and C_p criteria to select a subset of the nine candidate variables.

In the program in Figure 11.11, the quadratic and cross-product terms in the three explanatory variables are defined in the data step and then all nine terms are included as independent variables in the MODEL statement of the REGression procedure. By default, the analysis that would be produced is the one for the nine-predictor model. SAS makes provision for alternative analyses, however, by using the SELECTION= option in the model statement. One of the alternative analyses available is *all subsets regression*, which is called the RSQUARE selection procedure.

Figure 11.11. Program to run the RSQUARE selection option in REG

```
options ls=70 ps=54;
/*RSQUARE*/
data a;
   infile 'torrie.dat';
   input nitrogen chlorine potas logtime;
       nsq=nitrogen**2; csq=chlorine**2;
       psq=potas**2; nxc=nitrogen*chlorine;
       nxp=nitrogen*potas; cxp=chlorine*potas;
run;
proc reg data=a outest=est;
   model logtime=nitrogen chlorine potas nsq
       csq psq nxc nxp cxp/selection=rsquare
       cp;
run;
proc print data=est;
run;
proc plot data=est;
   plot _rsq_*_in_;
   plot _cp_*_p_='C' _p_*_p_='P'/overlay;
run;
```

The RSQUARE selection option performs, in this case, all $2^9 - 1 = 511$ possible regressions using some or all of the nine independent variables. Simply listing all 511 subsets takes 11 pages of printout! Therefore, only the portion of the output which pertains to three-predictor models is shown in Figure 11.12. A portion of Figure 11.12 is given in Rao's Example 11.24.

Figure 11.12. Three-predictor models from the RSQUARE selection option

```
   In   R-square        C(p)  Variables in Model

    3   0.8319402     3.5483  NITROGEN CHLORINE NXP
    3   0.8228892     4.9242  NITROGEN CHLORINE POTAS
    3   0.8215970     5.1207  CHLORINE NSQ NXP
    3   0.8200758     5.3519  NITROGEN CHLORINE PSQ
    3   0.8199244     5.3749  NITROGEN NXP CXP
    3   0.8150432     6.1170  CHLORINE POTAS NSQ
    3   0.8099406     6.8927  NSQ NXP CXP
    3   0.8097523     6.9213  CHLORINE NSQ PSQ
    3   0.8015408     8.1696  NITROGEN PSQ CXP
    3   0.8013364     8.2007  NITROGEN NXC NXP
    3   0.8001125     8.3867  NITROGEN POTAS CXP
    3   0.7965469     8.9288  NITROGEN CSQ NXP
    3   0.7923063     9.5734  POTAS NSQ CXP
    3   0.7907421     9.8112  NSQ NXC NXP
    3   0.7906464     9.8258  NSQ PSQ CXP
    3   0.7905777     9.8362  CHLORINE POTAS NXP
    3   0.7880435    10.2214  NITROGEN POTAS NXC
    3   0.7865634    10.4465  NITROGEN PSQ NXC
    3   0.7863145    10.4843  NSQ CSQ NXP
    3   0.7824056    11.0785  NITROGEN CHLORINE CXP
    3   0.7810887    11.2787  NITROGEN POTAS CSQ
    3   0.7804738    11.3722  POTAS NSQ NXC
    3   0.7773824    11.8421  NITROGEN CSQ PSQ
    3   0.7770743    11.8890  NSQ PSQ NXC
    3   0.7727320    12.5491  POTAS NSQ CSQ
    3   0.7704901    12.8899  CHLORINE NSQ CXP
    3   0.7682659    13.2280  CHLORINE PSQ NXP
    3   0.7667214    13.4628  NSQ CSQ PSQ
    3   0.7563587    15.0381  POTAS NXP CXP
    3   0.7561563    15.0689  POTAS NXC NXP
    3   0.7423985    17.1604  PSQ NXP CXP
    3   0.7421393    17.1998  PSQ NXC NXP
    3   0.7350080    18.2839  POTAS CSQ NXP
    3   0.7091521    22.2145  CSQ PSQ NXP
    3   0.7023277    23.2519  NITROGEN NXC CXP
    3   0.6953195    24.3173  NITROGEN CSQ CXP
    3   0.6941001    24.5026  NSQ NXC CXP
    3   0.6832976    26.1448  NSQ CSQ CXP
    3   0.6811588    26.4700  NITROGEN CHLORINE NXC
    3   0.6596724    29.7363  CHLORINE NSQ NXC
    3   0.6531281    30.7312  NITROGEN CHLORINE NSQ
    3   0.6429732    32.2749  NITROGEN CHLORINE CSQ
    3   0.6399118    32.7403  NITROGEN NSQ CSQ
    3   0.6326513    33.8440  NITROGEN NSQ NXC
    3   0.6305129    34.1691  NITROGEN CSQ NXC
    3   0.6272549    34.6644  CHLORINE NSQ CSQ
```

Note that subsets are grouped by the number of predictors in the model (denoted by "IN"), and arranged in descending order of R^2 in each group.

The default output for this option is to list all of the subsets and the value of R^2) for each. In the program in Figure 11.11, it is requested that the value of the C_p statistic be printed for each subset, as well. Although only the values of R^2 and C_p are shown in Figure 11.12, SAS has computed much more additional information about each model. This information is placed in an output SAS data set which is named EST in Figure 11.11. Using the command OUTEST=EST in the PROC REG command makes these statistics available for use in other SAS procedures. A small portion of the EST data set is shown in Figure 11.13; the entire data set takes 37 pages to print out for the 511 different subsets in this example.

The portion of the OUTEST=EST data set shown in Figure 11.13 was arbitrarily chosen for presentation. It shows information for subsets of independent variables numbered 183-196. These are all models containing four independent variables, as seen under the heading _IN_. Note that the name of the variable that denotes the number of independent variables in the model is

_IN_

and that the underlines are part of the variable name. Similarly, the number of parameters in the model is denoted by a variable named

_P_.

Note also that SAS's notation and Rao's notation are not consistent with each other: Rao uses p to denote the number of independent variables and p + 1 to stand for the number of parameters.

Figure 11.13. A portion of the OUTEST=EST data set

OBS	_MODEL_	_TYPE_	_DEPVAR_	_RMSE_	INTERCEP	NITROGEN	CHLORINE
183	MODEL1	PARMS	LOGTIME	0.23189	2.36240	.	.
184	MODEL1	PARMS	LOGTIME	0.23328	2.80890	-0.90843	.
185	MODEL1	PARMS	LOGTIME	0.23398	1.88921	.	.
186	MODEL1	PARMS	LOGTIME	0.23446	1.28417	.	.
187	MODEL1	PARMS	LOGTIME	0.23455	1.26614	.	.
188	MODEL1	PARMS	LOGTIME	0.23505	1.17577	.	.
189	MODEL1	PARMS	LOGTIME	0.23539	0.52265	.	.
190	MODEL1	PARMS	LOGTIME	0.23543	0.78349	.	.
191	MODEL1	PARMS	LOGTIME	0.23565	0.55135	.	.
192	MODEL1	PARMS	LOGTIME	0.23624	3.01975	-1.06541	.
193	MODEL1	PARMS	LOGTIME	0.23638	1.20346	.	.
194	MODEL1	PARMS	LOGTIME	0.23666	-0.03314	.	-0.44043
195	MODEL1	PARMS	LOGTIME	0.23669	2.30409	-0.90207	.
196	MODEL1	PARMS	LOGTIME	0.23698	1.60918	-0.47791	.

OBS	POTAS	NSQ	CSQ	PSQ	NXC	NXP
183	-0.44731	-0.21682	-0.19596	.	.	0.19619
184	.	0.01207	-0.17589	.	.	0.05869
185	-0.26506	-0.16496	.	.	-0.12971	0.14435
186	.	-0.14777	.	-0.019546	-0.12600	0.11926
187	.	-0.11442	-0.07609	.	-0.07700	0.06356
188	.	-0.07695	0.09007	0.035050	.	.
189	0.30038	-0.07732	0.03888	.	.	.
190	0.18886	-0.07752	.	0.010246	.	.
191	0.29217	-0.07938	.	.	0.01652	.
192	.	0.09059	.	0.021044	-0.11883	.
193	.	-0.08092	.	0.032392	0.03105	.
194	0.59789	.	.	-0.004433	.	-0.10492
195	0.19776	0.06656	.	.	-0.11698	.
196	0.19843	.	-0.05720	.	-0.08169	.

OBS	CXP	LOGTIME	_IN_	_P_	_EDF_	_RSQ_	_CP_
183	.	-1	4	5	25	0.79904	10.5500
184	.	-1	4	5	25	0.79663	10.9162
185	.	-1	4	5	25	0.79540	11.1027
186	.	-1	4	5	25	0.79455	11.2318
187	.	-1	4	5	25	0.79441	11.2538
188	-0.13099	-1	4	5	25	0.79353	11.3871
189	-0.10392	-1	4	5	25	0.79293	11.4786
190	-0.08743	-1	4	5	25	0.79287	11.4883
191	-0.09733	-1	4	5	25	0.79247	11.5486
192	.	-1	4	5	25	0.79143	11.7069
193	-0.11032	-1	4	5	25	0.79119	11.7434
194	.	-1	4	5	25	0.79069	11.8195
195	.	-1	4	5	25	0.79064	11.8269
196	.	-1	4	5	25	0.79013	11.9050

In examining the information in Figure 11.13 in a systematic fashion, after the subgroup number we see a column labeled _MODEL_. Since there was only one model statement used under PROC REG, _MODEL_ is always MODEL1 in this example. Similarly, _TYPE_ is always PARMS and _DEPVAR_ (dependent variable) is always LOGTIME. PARMS simply denotes that the parameter estimates are given for each subset of predictors. _RMSE_ is the square root of the error mean square for that set of predictors.

Then follow the estimated intercept and regression coefficients for each of the independent variables in that model. If a variable is not in the model under consideration, a dot is used in place of a coefficient. Since LOGTIME is the *dependent* variable, its coefficient is always 1.

As mentioned above, _IN_ is the number of independent variables in the subset, and _P_ = _IN_ + 1 is the number of parameters estimated (one for each predictor variable, plus an intercept). _EDF_ stands for error degrees of freedom, _RSQ_ stands for R^2, and _CP_ denotes C_p.

For example, Subset 183 uses the equation

$$\hat{y} = 2.36240 + 0.44731(\text{potas}) - 0.21682(\text{nsq}) - 0.19596(\text{csq}) + 0.19619(\text{nxp})$$

to estimate log leafburn time. There are four predictors, and thus five parameter estimates made. This equation explains R^2 = 79.904% of the variation in log leafburn time, the standard deviation of the estimated values from the actual values, based on $30 - (4 + 1) = 25$ degrees of freedom, is 0.23189, and the value of the C_p statistic is C_p = 10.5500.

Since one cannot realistically expect to be able to look at 511 pieces of information simultaneously in order to decide what subsets of independent variables are effective in explaining variation in log leafburn time, R^2-values are plotted against the number of predictors in the model in Figure 11.14, and the Cp-values are plotted against the number of parameters in the model in Figure 11.15. (Compare Figure 11.15 to Rao's Figure 11.11.)

Figure 11.14. Plot of R^2 vs. the number of independent variables

```
          Plot of _RSQ_*_IN_.  Legend: A = 1 obs, B = 2 obs, etc.
       |
       |
  1.0  +
       |
       |
       |
       |                          E       T       Z       U       G       A
       |               F          Z       Z       Z       N       B
  0.8  +               M          Z       Z       O       A
       |               I          O       G       A
       |               E          F       C
       |               B          C
       |               D          C       A
       |               C          B       A
R      |          E    G          I       E       A
- 0.6  +          B    K          H       A
s      |          D    E          B
q      |          H    D
u      |  A
a      |  A       A
r      |               A          A
e      |          B    F          D       A
d 0.4  +          D    D          A
       |          B    C
       |  A       B
       |
       |
       |  A       C    A
       |  A       A
  0.2  +
       |
       |  A       A
       |
       |  A
       |          A
       |  B
  0.0  +
       |
       ---+------+------+------+------+------+------+------+------+------+--
          1      2      3      4      5      6      7      8      9

                    Number of regressors in model
NOTE: 65 obs hidden.
```

Figure 11.15. Plot of C_p vs. the number of parameters

```
Plot of _CP_*_P_.  Symbol used is 'C'.
               Plot of _P_*_P_.   Symbol used is 'P'.

          |
   121.66 + C
          |         C
          |
          | C
          |
   106.66 + C       C
          |
          |
          |
          |
    91.66 + C       C       C
          | C       C
          |
M         |
a  76.66 +         C       C
l         | C       C       C
l         |         C       C       C
o         |         C       C       C       C
w         |         C       C       C
s  61.66 +         C       C
          |
C         |
(         | C       C
p         |
)  46.66 + C       C       C       C
          |         C       C       C       C
          |         C       C       C       C       C
          |                 C       C       C
          |         C       C       C
    31.66 +         C       C       C
          |                 C       C
          |                 C       C
          |                 C       C       C
          |                 C       C       C       C
    16.66 +                 C       C
          |                 C       C       C       C       C       C
          |                 C       C       C       C       C       C       C
          |                 C       C       C       C       C       C
          |                 C       C       C       C       C
     1.66 + P       P       C
        --+------+------+------+------+------+------+------+------+------+--
          2      3      4      5      6      7      8      9      10

                    Number of parameters in model

NOTE: 929 obs hidden.
```

In Figure 11.14, the value of R^2 increases quite a bit from the "best" single-predictor subset to the best two-predictor subset and then to the best three-predictor subset. There is a considerably smaller increase moving from the best three- to the best four-predictor subset, and then R^2 has "maxed out." At this point, we would consider some of the better three- and four-predictor models as possibilities.

Since there are so many points to be plotted in Figure 11.15 that many points are obscured, it is difficult to tell from the plot of C_p vs. the number of parameters that are useful models to consider. This plot would be more useful if there were not so many possible subsets. Referring to the C_p-values in Figure 11.12, we see that only one three-variable subset has $C_p <$ 4. While this is a reasonable candidate from the standpoint of the R^2 and C_p criteria, it is difficult to interpret since it contains the variables nitrogen and chlorine and the interaction between nitrogen and potassium but does not contain the variable, potassium. An examination of a few of the top four-predictor models, shown in Figure 11.16, indicates that using the three explanatory variables--nitrogen, chlorine, and potassium--along with the interaction between nitrogen and chlorine, gives a model with "good" statistics ($R^2 = 0.86$, $C_p = 1.6 < 5$), and also is easy to interpret and likely to make sense to the biochemist whose data this is. A multiple regression analysis of this model, along with partial regression significance tests, would be the next step in the analysis.

Figure 11.16. Some of the top four-predictor models from the RSQUARE selection

```
   In   R-square      C(p) Variables in Model

    4   0.8575449   1.6559 NITROGEN CHLORINE POTAS NXC
    4   0.8560214   1.8875 NITROGEN CHLORINE NXC NXP
    4   0.8501689   2.7772 NITROGEN CHLORINE PSQ NXC
    4   0.8472486   3.2211 CHLORINE POTAS NSQ NXC
    4   0.8440113   3.7133 CHLORINE NSQ NXC NXP
    4   0.8365307   4.8505 CHLORINE NSQ PSQ NXC
    4   0.8344633   5.1647 CHLORINE POTAS NXC NXP
    4   0.8343705   5.1789 NITROGEN CHLORINE CSQ NXP
    4   0.8338936   5.2514 NITROGEN CHLORINE POTAS NXP
```

12

The General Linear Model

12.1 Introduction

The General Linear Models (GLM) procedure in SAS is designed to handle a wide variety of quantitative and qualitative independent variables. GLM can be used to analyze analysis of variance models, in which all independent variables are qualitative; regression models in which all independent variables are quantitative; analysis of covariance models in which there are both qualitative and quantitative independent variables; and more. It works by adopting a dummy coding for qualitative variables, as explained in Rao's Chapter 12, and then regressing the quantitative dependent variable on the coded values of the independent variables.

12.2 A Regression Format for the ANOVA Model

Example 12.1. In an experiment to compare the effects of three therapies (treatments) for improving mental capacity, 30 subjects were randomly divided into three groups of 10; the three treatments were randomly assigned to the three groups. Each subject was given a test (pretest) requiring mental addition prior to his or her assigned treatment and another test (posttest) after the treatment was complete. The pretest (Z) and posttest (Y) scores for the 30 experimental subjects are as follows...

As always, the first step in the analysis of these data is to select an appropriate statistical model. Two identifiable factors may affect the response (posttest score): X, the type of treatment received by the subject (a qualitative factor); and Z, the subject's pretest score (a quantitative factor). Neither the ANOVA models nor the multiple linear regression models are equipped to handle qualitative and quantitative independent variables simultaneously, and so they are not suitable for analyzing the posttest scores data. What we need is a model that can handle qualitative and quantitative variables at the same time. The general linear is such a model and its use to analyze the posttest scores will be discussed in detail in the next section. In this example, we will ignore the effects of the pretest scores and illustrate how the ANOVA model for relating posttest scores to the treatment types can be expressed as a general linear model.

Figure 12.1. ANOVA procedure used to analyze posttest scores as a function of treatment group

```
options ls=70 ps=54;
/*Pretest-Posttest in 3 Treatments*/
data a;
 input tmt @@;
   do i = 1 to 10;
     input pre post @@;
     output;
   end;
lines;
1 24 45 28 50 38 59 42 60 24 47 39 66 45 76
       19 50 19 39 22 36
2 23 28 33 39 31 36 34 39 18 22 24 28 41 49
       34 39 30 33 39 43
3 27 34 27 31 44 55 38 43 32 44 26 28 24 33
       13 13 36 39 52 58
;
run;
proc anova;
  class tmt;
  model post=tmt;
  means tmt;
  means tmt/lsd cldiff;
run;
```

The ANOVA procedure could be used to analyze the posttest scores as a function of treatment group, ignoring pretest scores. Figure 12.1 shows the program to perform this analysis, and the output is shown in Figures 12.2

Figure 12.2. **ANOVA analysis of posttest scores as a function of treatment group**

```
                  Analysis of Variance Procedure
                     Class Level Information

                 Class     Levels     Values

                  TMT         3        1 2 3

            Number of observations in data set = 30


Dependent Variable: POST

Source                  DF    Sum of Squares    F Value     Pr > F

Model                    2     1752.26666667      6.71      0.0043

Error                   27     3527.60000000

Corrected Total         29     5279.86666667

                R-Square               C.V.            POST Mean

                0.331877            27.17188          42.0666667


Source                  DF       Anova SS       F Value     Pr > F

TMT                      2     1752.26666667      6.71      0.0043
```

and 12.3. Examine the output in Figure 12.3 to note the treatment means, standard errors, and the estimated differences between treatment means.

Now suppose that the treatments are given the following dummy coding:

$$X_1 = 1 \text{ if Treatment} = 1 \qquad\qquad X_2 = 1 \text{ if Treatment} = 2$$
$$ 0 \text{ otherwise} \qquad\qquad\qquad 0 \text{ otherwise}$$

and that then the posttest scores are regressed on X_1 and X_2. The SAS program to perform this analysis is shown in Figure 12.4, and the resulting output is given in Figure 12.5.

Figure 12.3. Analysis of treatment means from ANOVA procedure

```
                    Analysis of Variance Procedure

      Level of               ------------POST------------
      TMT           N        Mean                SD

       1           10      52.8000000        12.4078828
       2           10      35.6000000         8.0027773
       3           10      37.8000000        13.1892212

              T tests (LSD) for variable: POST

  NOTE: This test controls the type I comparisonwise error rate
        not the experimentwise error rate.

      Alpha= 0.05  Confidence= 0.95  df= 27  MSE= 130.6519
                  Critical Value of T= 2.05183
              Least Significant Difference= 10.489

  Comparisons significant at the 0.05 level are indicated by '***'.

                       Lower      Difference      Upper
               TMT     Confidence   Between    Confidence
           Comparison    Limit      Means        Limit

       1    - 3         4.511      15.000       25.489    ***
       1    - 2         6.711      17.200       27.689    ***

       3    - 1       -25.489     -15.000       -4.511    ***
       3    - 2        -8.289       2.200       12.689

       2    - 1       -27.689     -17.200       -6.711    ***
       2    - 3       -12.689      -2.200        8.289
```

Compare the analysis of variance tables in Figures 12.2 and 12.5: they are identical. Now compare the estimates of treatment means at the top part of Figure 12.3 with the parameter estimates in Figure 12.5. Recall that the dummy coding adopted for the regression analysis used Treatment 3 as the reference group. The mean for Treatment 3, given as 37.8 in Figure 12.3, is found as the estimate of the INTERCEP(T) in Figure 12.5:

$$b_0 = \bar{y}_3.$$

The estimate of the coefficient on X_1 in Figure 12.5 can be seen to be the *difference* in the mean for Treatment 1 and the mean for Treatment 3:

Figure 12.4. Program to analyze posttest scores using REG

```
options ls=70 ps=54;
/*Pretest-Posttest in 3 Treatments*/
data a;
 input tmt @@;
   do i = 1 to 10;
      input pre post @@;
      output;
   end;
lines;
1 24 45 28 50 38 59 42 60 24 47 39 66 45 76
        19 50 19 39 22 36
2 23 28 33 39 31 36 34 39 18 22 24 28 41 49
        34 39 30 33 39 43
3 27 34 27 31 44 55 38 43 32 44 26 28 24 33
        13 13 36 39 52 58
;
run;
data b; set a;
if tmt = 1 then x1 = 1;
   else x1 = 0;
if tmt = 2 then x2 = 1;
   else x2 = 0;
run;
proc reg data=b;
   model post=x1 x2;
run;
```

$$b_0 + b_1 = \bar{y}_1$$

or

$$37.8 + 15.0 = 52.8.$$

Similarly, the coefficient on X_2 is the difference in the means for Treatments 2 and 3:

$$b_0 + b_2 = \bar{y}_2$$

or

$$37.8 - 2.2 = 35.6.$$

Figure 12.5. Analysis of posttest scores using REG

```
Dependent Variable: POST

                        Analysis of Variance

                       Sum of          Mean
Source          DF     Squares         Square      F Value      Prob>F

Model            2    1752.26667     876.13333       6.706      0.0043
Error           27    3527.60000     130.65185
C Total         29    5279.86667

       Root MSE        11.43030      R-square       0.3319
       Dep Mean        42.06667      Adj R-sq       0.2824
       C.V.            27.17188

                        Parameter Estimates

                    Parameter      Standard     T for H0:
   Variable   DF     Estimate        Error     Parameter=0    Prob > |T|

   INTERCEP    1    37.800000     3.61457953      10.458        0.0001
   X1          1    15.000000     5.11178739       2.934        0.0067
   X2          1    -2.200000     5.11178739      -0.430        0.6703
```

These relations can be further verified by looking at the Differences Between Means in the bottom portion of Figure 12.3.

We have demonstrated that a one-way analysis of variance can be analyzed as a regression on dummy variables. The ANOVA procedure in fact performs a multiple regression analysis, but has a *built-in* facility for converting the qualitative variables to numerical dummy codings: the CLASS statement.

Figure 12.6 shows the same data analyzed using the GLM procedure. Note that, as in ANOVA, using the CLASS statement in GLM causes the categorical variable to be converted to a dummy coding. The analysis of variance portion of Figure 12. 7 is identical to the analysis of variance tables in Figures 12.2 and 12.5. Without the

```
/solution
```

Figure 12.6. Program for analyzing posttest scores as a function of treatment group using the GLM procedure

```
options ls=70 ps=54;
/*Pretest-Posttest in 3 Treatments*/
data a;
 input tmt @@;
   do i = 1 to 10;
     input pre post @@;
     output;
   end;
lines;
1 24 45 28 50 38 59 42 60 24 47 39 66 45 76
        19 50 19 39 22 36
2 23 28 33 39 31 36 34 39 18 22 24 28 41 49
        34 39 30 33 39 43
3 27 34 27 31 44 55 38 43 32 44 26 28 24 33
        13 13 36 39 52 58
;
run;
proc glm;
  class tmt;
  model post=tmt/solution;
 run;
```

option in the MODEL statement, the printout in Figure 12.7 would have ended with the Type I and Type III analyses of the treatment factor. Adding this command produces the parameter estimates at the bottom of the figure. Comparing this portion of the output from GLM with the parameter estimates from REG in Figure 12.5, we see that they are essentially the same, except for labeling. The estimate for Treatment 3 being given as zero indicates that Treatment 3 is being used as the reference group. *By default, SAS will use the group with the highest number or the latest letter in the alphabet as the reference group. If the data being analyzed has a natural reference group, say a control condition, then take care to name or number it so that it will be used as the reference group by the GLM procedure.*

Figure 12.7. GLM analysis of posttest scores

```
                    General Linear Models Procedure
                       Class Level Information

                 Class      Levels    Values

                  TMT          3       1 2 3


            Number of observations in data set = 30
```

Dependent Variable: POST

Source	DF	Sum of Squares	F Value	Pr > F
Model	2	1752.26666667	6.71	0.0043
Error	27	3527.60000000		
Corrected Total	29	5279.86666667		

R-Square	C.V.	POST Mean
0.331877	27.17188	42.0666667

Source	DF	Type I SS	F Value	Pr > F
TMT	2	1752.26666667	6.71	0.0043

Source	DF	Type III SS	F Value	Pr > F
TMT	2	1752.26666667	6.71	0.0043

Parameter		Estimate	T for H0: Parameter=0	Pr > |T|	Std Error of Estimate
INTERCEPT		37.80000000 B	10.46	0.0001	3.61457953
TMT	1	15.00000000 B	2.93	0.0067	5.11178739
	2	-2.20000000 B	-0.43	0.6703	5.11178739
	3	0.00000000 B	.	.	.

```
NOTE: The X'X matrix has been found to be singular and a generalized
      inverse was used to solve the normal equations.   Estimates
      followed by the letter 'B' are biased, and are not unique
      estimators of the parameters.
```

Finally, note at the bottom of Figure 12.7 the message:

```
NOTE: The X'X matrix has been found to be singular
and a generalized inverse was used to solve the
normal equations.  Estimates followed by the letter
'B' are biased, and are not unique estimators of
the parameters.
```

For some rather technical reasons having to do with being a very general purpose procedure, GLM fits the model shown in Rao's Equation 12.9; that is, one with t, instead of $t - 1$, dummy variables to indicate t levels of the categorical variable. Thus, the GLM model is "overparameterized", and there is no unique solution to the normal equations which need to be solved in order to obtain the parameter estimates, as explained by Rao in Section 12.2. The solution used by GLM has the parameter estimate for Treatment 3 to be zero, as seen in the GLM output in Figure 12.7. You need not worry about parameter estimates being biased in any GLM analyses presented here. Should it be of interest to examine the $\mathbf{X'X}$ and $\mathbf{S}$ matrices, GLM will produce these if the options "xpx" and "i" (for inverse), respectively, are included in the model statement. Writing

```
model post=tmt/solution xpx i;
```

in the program in Figure 12.6 will give, in addition to the output in Figure 12.7, the matrices shown in Figures 12.8 and 12.9. (The rows and columns labeled POST in Figures 12.8 and 12.9 have to do with the dependent variable, Y, and are not part of the $\mathbf{X'X}$ or $\mathbf{S}$ matrices.)

In Figure 12.7, Treatment 3 is the reference group and the estimate of the "intercept", 37.8, is the mean response in Treatment 3. GLM gives a built-in test of whether this mean differs significantly from zero: $t = 10.46$ with $p = 0.0001$ says that this mean does differ from zero. The estimate for Treatment 1, 14.9, is the difference between the means of Treatments 1 and 3; $t = 2.923$ with $p = 0.0067$ says that the means of Treatments 1 and 3 differ more than might be expected by chance alone. On the other hand, the mean posttest scores in Treatments 2 and 3 do not differ significantly: $t = -0.43$, p

Figure 12.8. X'X matrix for posttest scores as a function of treatment

```
                        The X'X Matrix

             INTERCEPT        TMT 1        TMT 2        TMT 3

INTERCEPT           30           10           10           10
TMT 1               10           10            0            0
TMT 2               10            0           10            0
TMT 3               10            0            0           10
POST              1262          528          356          378


             POST

INTERCEPT         1262
TMT 1              528
TMT 2              356
TMT 3              378
POST             58368
```

Figure 12.9. S = (X'X)$^{-1}$ matrix for posttest scores as a function of treatment

```
                  X'X Generalized Inverse (g2)

             INTERCEPT        TMT 1        TMT 2        TMT 3

INTERCEPT          0.1         -0.1         -0.1            0
TMT 1             -0.1          0.2          0.1            0
TMT 2             -0.1          0.1          0.2            0
TMT 3                0            0            0            0
POST              37.8           15         -2.2            0


             POST

INTERCEPT         37.8
TMT 1               15
TMT 2             -2.2
TMT 3                0
POST            3527.6
```

= 0.6703. There are, however, other questions that might be of interest: are the means of Treatments 1 and 2 significantly different from zero (or some other value)? Do the means of Treatments 1 and 2 differ significantly from each other? As illustrated by Rao in Section 12.2, the **S** matrix can be used to compute the estimated standard error of any linear combination of the regression coefficients, so that confidence intervals can be constructed and/or tests of hypotheses can be performed. It is possible to get SAS to perform other tests of linear combinations of regression coefficients, by using the TEST command in the REG procedure--*not the GLM procedure.* (There is a TEST command in GLM, but it does something altogether different, which we shall see in Chapter 14: Random Effects Models.)

Refer back to the program for the REG analysis in Figure 12.4, and the output in Figure 12.5. Suppose you want to perform a test of whether the means for Treatments 1 and 2 differ significantly. The coefficient on X_1 measures the difference between means of Treatments 1 and 3,

$$b_1 = \bar{y}_1 - \bar{y}_3,$$

and the coefficient on X_2 measures the difference between means of Treatments 2 and 3,

$$b_2 = \bar{y}_2 - \bar{y}_3.$$

It is easily seen that the difference in the regression coefficients estimates the difference in the means of Treatments 1 and 2:

$$b_1 - b_2 = (\bar{y}_1 - \bar{y}_3) - (\bar{y}_2 - \bar{y}_3)$$

$$= \bar{y}_1 - \bar{y}_2.$$

To perform the test of $b_1 - b_2$ in REG, add the following commands after the MODEL statement in Figure 12.4:

```
t1vs2: test x1-x2;
```

The "t1vs2" followed by a colon is simply a label and does not give instructions to the computer. Whatever label you attach here will be used to label the printout resulting from this test. Since you might want to test several different hypotheses, using labels allows you to identify what results go with what test. (Note that REG objects if the label begins with a number instead of with a letter.) The command, "test x1–x2" really means to test the difference in the *coefficients* on X_1 and X_2. It is understood that "test x1–

x2" means to test whether the difference in the coefficients is *zero*; the command could have been written

```
t1vs2: test x1-x2=0;
```

The implication is that you can test whether the difference is *any* hypothesized value. Suppose, for example, that there is reason to believe that the mean posttest score in Treatment 1 should be at least 10 points higher than the mean in Treatment 2. Writing

```
t1vs2: test x1-x2=10;
```

will perform this test.

If it is of interest to test hypotheses concerning individual treatment means, instead of their differences, the TEST command can be used to do this as well. The mean of Treatment 3 is estimated by the intercept. What combinations of regression coefficients estimate the means of Treatments 1 and 2? A little algebra shows that

$$b_1 + b_0 = (\bar{y}_1 - \bar{y}_3) + \bar{y}_3 = \bar{y}_1$$

and

$$b_2 + b_0 = (\bar{y}_2 - \bar{y}_3) + \bar{y}_3 = \bar{y}_2.$$

As an example, suppose that we want to test the following hypotheses:

$$H_0: \mu_1 = 50$$
$$H_0: \mu_2 = 30$$
$$H_0: \mu_3 = 40$$
$$H_0: \mu_1 - \mu_2 = 10.$$

The following TEST commands, added to the program in Figure 12.4 after the MODEL statement, produced the output shown in Figure 12.10.

```
tmt1: test x1+intercept=50;
```

```
tmt2: test x2+intercept=30;
  tmt3: test intercept=40;
    t1vs2: test x1-x2=10;
```

The results in Figure 12.10 indicate that none of the four hypotheses will be rejected at any reasonable level.

Figure 12.10. TEST commands in the REG procedure

```
Dependent Variable: POST
Test: TMT1      Numerator:      78.4000  DF:   1   F value:    0.6001
                Denominator:   130.6519  DF:  27   Prob>F:     0.4453


Dependent Variable: POST
Test: TMT2      Numerator:     313.6000  DF:   1   F value:    2.4003
                Denominator:   130.6519  DF:  27   Prob>F:     0.1330


Dependent Variable: POST
Test: TMT3      Numerator:      48.4000  DF:   1   F value:    0.3705
                Denominator:   130.6519  DF:  27   Prob>F:     0.5478


Dependent Variable: POST
Test: T1VS2     Numerator:     259.2000  DF:   1   F value:    1.9839
                Denominator:   130.6519  DF:  27   Prob>F:     0.1704
```

12.3 Models with Nominal and Interval Variables

In order to incorporate information concerning pretest scores into the analysis of posttest scores as a function of treatment group, all that is necessary is to change the MODEL statement in the program in Figure 12.6 to read:

```
model post=tmt pre/solution I;
```

Declaring TMT to be a CLASS variable in GLM will result in the three treatment levels being converted to dummy variables. The variable PRE is treated as a numerical variable, and, essentially, a multiple regression is performed relating posttest scores to the dummy variables and the pretest scores.

Figure 12.11. Class levels information and S matrix for posttest scores as a function of treatment and pretest scores

```
                General Linear Models Procedure
                   Class Level Information

             Class     Levels     Values

             TMT            3      1 2 3


        Number of observations in data set = 30

            X'X Generalized Inverse (g2)

             INTERCEPT          TMT 1           TMT 2       TMT 3

INTERCEPT    0.5118211251 -0.124528531 -0.115491704            0
TMT 1       -0.124528531   0.201460947 0.1009227034            0
TMT 2       -0.115491704  0.1009227034    0.20058276           0
TMT 3                  0             0             0            0
PRE         -0.012909753 0.0007689195 0.0004856333             0
POST        1.7998624039 17.144208822 -0.845762849             0


             PRE           POST

INTERCEPT   -0.012909753  1.7998624039
TMT 1        0.0007689195 17.144208822
TMT 2        0.0004856333 -0.845762849
TMT 3                  0             0
PRE          0.0004046945 1.1285309591
POST         1.1285309591 380.57856738
```

The Class Levels Information and the **S** matrix are given in Figure 12.11. In Figure 12.12, the Model sum of squares with 3 degrees of freedom is broken down into its components: TMT with 2 degrees of freedom, and PRE with 1 degree of freedom. This decomposition is given as either a Type I (sequential) or as a Type III (partial) sum of squares. The difference in the Type I and Type III sums of squares for TMT is that the Type III is adjusted for pretest scores, while the Type I is not. The analysis indicates that both the treatment groups and the pretest scores affected posttest scores.

From the Parameter Estimates portion of Figure 12.12, the regression equation is:

Figure 12.12. GLM analysis of posttest scores as a function of treatment and pretest scores

```
                         General Linear Models Procedure

Dependent Variable: POST

Source                    DF      Sum of Squares      F Value      Pr > F

Model                      3      4899.28809929        111.57      0.0001

Error                     26       380.57856738

Corrected Total           29      5279.86666667

                     R-Square               C.V.              POST Mean

                     0.927919            9.094889            42.0666667


Source                    DF          Type I SS      F Value      Pr > F

TMT                        2      1752.26666667        59.85      0.0001
PRE                        1      3147.02143262       215.00      0.0001

Source                    DF        Type III SS      F Value      Pr > F

TMT                        2      2052.22695991        70.10      0.0001
PRE                        1      3147.02143262       215.00      0.0001


                                     T for H0:    Pr > |T|    Std Error of
Parameter            Estimate       Parameter=0                 Estimate

INTERCEPT          1.79986240 B          0.66      0.5166      2.73712476
TMT          1    17.14420882 B          9.98      0.0001      1.71723972
             2    -0.84576285 B         -0.49      0.6257      1.71349283
             3     0.00000000 B            .           .             .
PRE                1.12853096           14.66      0.0001      0.07696604

NOTE: The X'X matrix has been found to be singular and a generalized
      inverse was used to solve the normal equations.   Estimates
      followed by the letter 'B' are biased, and are not unique
      estimators of the parameters.
```

$$\hat{y} = 1.7999 + 17.1442(\text{tmt1}) - 0.8458(\text{tmt2}) + 0.0000(\text{tmt3}) + 1.1285(\text{pre})$$
$$= 1.7999 + 17.1442(\text{tmt1}) - 0.8458(\text{tmt2}) + 1.1285(\text{pre}).$$

For a subject receiving Treatment 1, the equation becomes

$$\hat{y} = 1.7999 + 17.1442(\text{tmt1}) - 0.8458(\text{tmt2}) + 0.0000(\text{tmt3}) + 1.1285(\text{pre})$$

$$= 1.7999 + 17.1442(\text{tmt1}) - 0.8458(\text{tmt2}) + 1.1285(\text{pre}).$$

For a subject receiving Treatment 1, the equation becomes

$$\hat{y} = 1.7999 + 17.1442(1) - 0.8458(0) + 1.1285(\text{pre})$$
$$= 18.9441 + 1.1285(\text{pre}).$$

The equation for a subject receiving Treatment 2 is

$$\hat{y} = 1.7999 + 17.1442(0) - 0.8458(1) + 1.1285(\text{pre})$$
$$= 0.9541 + 1.1285(\text{pre}),$$

and for a subject receiving Treatment 3 we have

$$\hat{y} = 1.7999 + 17.1442(0) - 0.8458(0) + 1.1285(\text{pre})$$
$$= 1.7999 + 1.1285(\text{pre}).$$

For subjects in each of the three treatment groups, posttest score is related to pretest score, as demonstrated by the p-value of 0.0001 associated with the partial t-test or partial F-test for the variable PRE. The p-value for the intercept, which is the intercept for Treatment 3, is not significantly different from zero, indicating that for Treatment 3, the average posttest score of subjects scoring zero on the pretest could also be zero. The p-value of 0.9892 indicates that there is no significant difference between Treatments 2 and 3, but $p = 0.0001$ for Treatment 1 indicates that this group does differ significantly from Treatment 3.

12.4 Modeling Interactions

Example 12.4. In an experiment to compare the uptake of a chemical from two dietary supplements, 14 animals were randomly divided into two groups of seven. The following data show the initial body weight (kg) and the chemical uptake (g/day) of the experimental animals...The measured values of the chemical uptake are plotted against the initial weights in

[Rao's] Figure 12.2. An examination of Figure 12.2 suggests that the relationship between the chemical uptake and initial weight can be approximated by two straight lines that are not parallel (of different slopes). The different slopes of the lines suggest that the difference between the expected responses for the two diets is not the same at different initial weights. Indeed, it appears that the differential response to the two diets increases with the initial weight. Thus, Figure 12.2 indicates an interaction between diet (a qualitative explanatory factor) and initial weight (a quantitative explanatory factor), because the change in the expected response that results from changing the type of diet appears to depend on the initial weight.

Figure 12.13 shows a plot of the data similar to Rao's Figure 12.2. It is obvious that uptake is greater under Diet 2 than under Diet 1, and it also appears that the slope relating uptake to initial weight is steeper under Diet 2 than under Diet 1. The SAS program in Figure 12.14 shows the code which produced the plot in Figure 12.13 and also the commands necessary to fit equations with different slopes to the data for the two diets, and test for the significance of the difference in the slopes. Figure 12.15 shows Class Levels Information and the **S** matrix. The GLM analysis is shown in Figure 12.16 and a plot of the data with predicted values is given in Figure 12.17. Most of Figures 12.15 and 12.16 is shown in Rao's analysis of Example 12.4.

Let us first inspect Figure 12.15. In addition to the usual Class Levels Information that resulted from declaring DIET to be a CLASS variable, there is also a section labeled "Matrix Element Representation" which came about by defining the interaction (cross-product) between DIET and WEIGHT in the MODEL statement. In this section, SAS assigns its own names to the qualitative and quantitative variables and their interaction. The interaction between weight and diet is designated as two dummy variables, "DUMMY001" and "DUMMY002". From the **S** matrix (see Part 4 of Rao's Example 12.4) we see by the zeroes in all rows and columns involving DIET 2 that Diet 2 is serving as the reference group.

Figure 12.13. Plot of uptake against initial weight for two diets

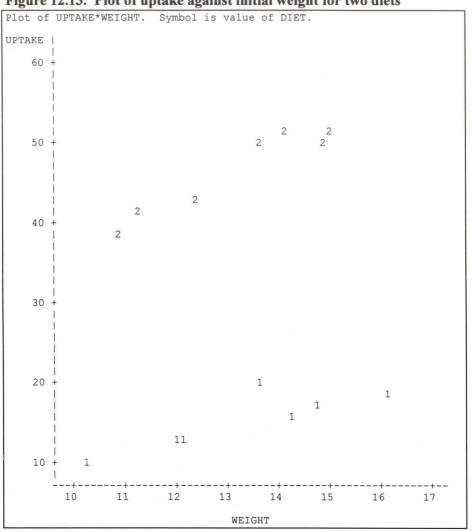

Figure 12.14. Program to fit equations with different slopes for the two diets

```
options ls=70 ps=54;
data supp;
  input diet @@;
    do i = 1 to 7;
      input weight uptake @@;
      output;
    end;
  drop i;
lines;
1 12.1 12.7 10.2 10.3 13.6 19.8 14.8 17.4
16.1 18.9 14.3 16.1 12.0 12.7
2 11.3 41.4 12.4 43.3 13.6 49.7 14.1 51.6
15.0 51.5 14.9 50.5 10.9 38.4
;
run;
proc plot;
  plot uptake*weight=diet;
run;
proc glm;
  class diet;
  model uptake = diet weight
      diet*weight/solution i;
  output out=pred p=yhat;
run;
proc plot data=pred;
  plot uptake*weight=diet
      yhat*weight='*'/overlay;
run;
```

Figure 12.16 shows the analysis of variance table, the Type I and Type III sums of squares and F-tests, and the parameter estimates, as in Parts 1-3 of Rao's Example 12.4. The significant WEIGHT*DIET interaction says that effects of diet need to be interpreted with reference to the initial weight. Under Parameter Estimates, the INTERCEPT and the coefficient on weight are the intercept and slope, respectively of the equation relating uptake to initial weight for animals on Diet 2 (the reference group):

Figure 12.15. Class levels information and S matrix for uptake as a function of initial weight and diet

```
General Linear Models Procedure
                    Class Level Information

                Class     Levels     Values

                DIET          2       1 2


            Number of observations in data set = 14


            Matrix Element Representation

Dependent Variable: UPTAKE

                Effect              Representation

                INTERCEPT           INTERCEPT

                DIET        1        DIET 1
                            2        DIET 2

                WEIGHT              WEIGHT

                WEIGHT*DIET 1        DUMMY001
                            2        DUMMY002


                X'X Generalized Inverse (g2)

              INTERCEPT        DIET 1        DIET 2         WEIGHT

INTERCEPT     10.57231192  -10.57231192          0  -0.791824115
DIET 1       -10.57231192   18.110235953         0   0.7918241154
DIET 2                 0             0            0             0
WEIGHT       -0.791824115   0.7918241154         0   0.0601167984
DUMMY001      0.7918241154 -1.347844182          0  -0.060116798
DUMMY002               0             0            0             0
UPTAKE        4.9978529715 -10.36759736          0   3.1606836139


              DUMMY001       DUMMY002        UPTAKE

INTERCEPT     0.7918241154          0   4.9978529715
DIET 1       -1.347844182           0  -10.36759736
DIET 2                 0            0             0
WEIGHT       -0.060116798           0   3.1606836139
DUMMY001      0.1019228184          0  -1.597974584
DUMMY002               0            0             0
UPTAKE       -1.597974584           0   31.093594967
```

Figure 12.16. GLM analysis of uptake as a function of initial weight and diet

```
                      General Linear Models Procedure

Dependent Variable: UPTAKE

Source                    DF    Sum of Squares    F Value      Pr > F

Model                      3      3634.74997646    389.66      0.0001

Error                     10        31.09359497

Corrected Total           13      3665.84357143

                  R-Square              C.V.             UPTAKE Mean

                  0.991518            5.684257             31.0214286


Source                    DF        Type I SS    F Value      Pr > F

DIET                       1      3410.16071429   1096.74      0.0001
WEIGHT                     1       199.53576768     64.17      0.0001
WEIGHT*DIET                1        25.05349450      8.06      0.0176

Source                    DF      Type III SS    F Value      Pr > F

DIET                       1         5.93515597      1.91      0.1972
WEIGHT                     1       218.89541925     70.40      0.0001
WEIGHT*DIET                1        25.05349450      8.06      0.0176


                                  T for H0:    Pr > |T|   Std Error of
Parameter            Estimate   Parameter=0                 Estimate

INTERCEPT          4.99785297 B       0.87      0.4038      5.73350839
DIET        1    -10.36759736 B      -1.38      0.1972      7.50408117
            2      0.00000000 B        .          .            .
WEIGHT             3.16068361 B       7.31      0.0001      0.43234794
WEIGHT*DIET 1     -1.59797458 B      -2.84      0.0176      0.56295176
            2      0.00000000 B        .          .            .

NOTE: The X'X matrix has been found to be singular and a generalized
      inverse was used to solve the normal equations.   Estimates
      followed by the letter 'B' are biased, and are not unique
      estimators of the parameters.
```

$$\hat{y} = 4.9979 + 3.1607 \text{ WEIGHT}.$$

The *p*-values associated with these two coefficients indicate that for this group, the intercept does not differ significantly from zero, but that animals with higher initial weight tend to have higher uptake.

The coefficients labeled DIET 1 and WEIGHT*DIET 1 are the differences in intercepts and slopes, respectively, in the two diets. Thus, for animals fed Diet 1, the equation relating uptake to initial weight is

$$\hat{y} = (4.9979 - 10.3676) + (3.1607 - 1.5980)\text{WEIGHT}$$
$$= -5.3697 + 1.5627 \text{ WEIGHT}.$$

The *p*-value provided for the coefficient for Diet 1 indicates that the intercepts of the two equations do not differ significantly; the *p*-value for the WEIGHT*DIET 1 coefficient indicates that the slopes of the two equations are not significantly different.

Figure 12.17 graphs the two equations relating uptake to initial weight.

How should we interpret the results of this analysis? We are told for Diet 2 that animals with heavier initial weight generally have higher uptake: specifically, for each additional one gram of weight, the uptake increases by 3.16 grams. For Diet 1, the slope is not so steep: for these animals, an additional one gram of initial weight corresponds to an additional

$$3.1606 - 1.5979 \approx 1.56$$

grams of uptake. Thus, the effect of diet on uptake is not that animals on one diet have higher or lower uptake than animals on the other diet (as shown by the nonsignificant Type III *F*-test on diet), but that the *rate of increase of uptake with initial weight* is higher for animals on Diet 2 than for those on Diet 1.

Figure 12.17. Data and predicted values for uptake as a function of initial weight and diet

```
          Plot of UPTAKE*WEIGHT.   Symbol is value of DIET.
          Plot of YHAT*WEIGHT.     Symbol used is '*'.

UPTAKE |
       |
    60 +
       |
       |
       |
       |
       |                                      2      *2
    50 +                               2      *       2
       |                               *
       |
       |                      *
       |                      2
       |              2
    40 +        *     *
       |        2
       |
       |
       |
    30 +
       |
       |
       |
       |
    20 +                              1                    *
       |                                          *        1
       |                              *     1     1
       |
       |          11
    10 +    1
       |
       ---+-------+-------+-------+-------+-------+-------+-------+-------+--
          10      11      12      13      14      15      16      17

                                    WEIGHT

NOTE: 5 obs hidden.
```

Figure 12.18. Program for using the TEST command in REG when slopes are unequal

```
options ls=70 ps=54;
data supp;
  input diet @@;
    do i = 1 to 7;
      input weight uptake @@;
      output;
    end;
  drop i;
lines;
1 12.1 12.7 10.2 10.3 13.6 19.8 14.8 17.4
16.1 18.9 14.3 16.1 12.0 12.7
2 11.3 41.4 12.4 43.3 13.6 49.7 14.1 51.6
15.0 51.5 14.9 50.5 10.9 38.4
run;
data b; set supp;
  if diet = 1 then x1=1;
    else x1=0;
  interact=x1*weight;
run;
proc reg;
  model uptake = x1 weight interact;
  a: test weight+interact;
run;
```

We know that this rate of increase is significantly different from zero for animals on Diet 2, and that the rate for Diet 2 is greater than the rate for Diet 1. What remains to be answered is whether the rate of increase for animals on Diet 1 is significantly different from zero. This question can be answered by using the TEST command in REG. Figure 12.18 shows a program to perform this test.

In Figure 12.18, a dummy variable indicating diet is defined in such a way that Diet 2 is the reference group. Additionally, a new variable, which is the interaction between the dummy indicator of diet and the quantitative variable, WEIGHT, is defined in a data step. In the REG procedure, the model is that uptake is a function of the group indicator, the initial weight, and the interaction between the two. Resulting parameter estimates in Figure 12.19 are the same as in Figure 12.16. The results of the TEST command indicate that although the rate of increase of uptake with initial weight is lower under Diet 1 than under Diet 2, it is still significantly greater than zero under Diet 1.

Figure 12.19. Analysis for unequal slopes using REG

```
                     Analysis of Variance

                       Sum of          Mean
Source         DF      Squares         Square      F Value      Prob>F

Model           3    3634.74998     1211.58333     389.657      0.0001
Error          10      31.09359        3.10936
C Total        13    3665.84357

       Root MSE          1.76334    R-square      0.9915
       Dep Mean         31.02143    Adj R-sq      0.9890
       C.V.              5.68426

                     Parameter Estimates

                   Parameter      Standard      T for H0:
Variable    DF      Estimate         Error    Parameter=0      Prob > |T|

INTERCEP     1      4.997853    5.73350839          0.872          0.4038
X1           1    -10.367597    7.50408117         -1.382          0.1972
WEIGHT       1      3.160684    0.43234794          7.311          0.0001
INTERACT     1     -1.597975    0.56295176         -2.839          0.0176


Dependent Variable: UPTAKE
Test: A          Numerator:      58.4141   DF:    1    F value:  18.7865
                 Denominator:   3.109359   DF:   10    Prob>F:    0.0015
```

Example 12.5. In an experiment to evaluate two minerals--sodium selenite and calcium selenite--as dietary sources of, selenium (Se) for broiler chickens, two diets--Diet A and Diet B--were formed by supplementing the standard basal diet with known amounts of sodium and calcium selenite, respectively. Among other things, the investigators were interested in studying how Y, the selenium concentration (μg/g) in the liver, was affected by the four experimental factors:

- X_1, the type of diet (1 for A and 0 for B);
- X_2, the sex of the animal (F or female and M or male);
- X_3, the concentration (μg/g) in the diet;
- X_4, the feeding period (weeks--that is, the time period over which the diet was fed to the animal.

The data collected in this study have been modified for illustration purposes--to ensure that certain interactions are significant .

Some questions that were of interest to the investigators are as follows:

1. Is there a difference between the liver Se concentration for the two diets?
2. Does the effect of diet on liver Se depend on the animal's gender? In other words, is there an interaction between diet and gender?
3. Does the effect of diet on liver Se depend on the feeding period? In other words, is there an interaction between diet and feeding period?
4. Is there an effect of the Se concentration in the diet? Is there an interaction between feeding period and dietary Se concentration?

Example 12.6. Let's analyze the liver Se concentration data described in Example 12.5. The first step is to select a model describing how the response variable Y = liver Se concentration is related to the four experimental factors: X_1, diet; X_2, gender; X_3, dietary Se concentration; and X_4, feeding period. In Section 11.10, we looked at methods of selecting a suitable subset from a set of candidate independent variables for inclusion in a multiple linear regression model. The same methods can be used to select the independent variables for a general linear model. For illustration purposes, let's assume that the general linear model with p = 10 independent variables,

$$Y = \beta_0 + \beta_1 x_1 + \beta_2 x_2 + \beta_3 x_3 + \beta_4 x_4$$
$$+ \beta_{12} x_1 x_2 + \beta_{13} x_1 x_3 + \beta_{14} x_1 x_4$$
$$+ \beta_{23} x_2 x_3 + \beta_{24} x_2 x_4 + \beta_{34} x_3 x_4 + E,$$

is selected for analysis of the liver Se concentration data.

Figure 12.20. SAS program to analyze selenium concentration data

```
options ls=70 ps=54;
data selenium;
  infile 'selenium.dat';
  do i = 1 to 3;
    input dietse @@;
      do j = 1 to 2;
          input diettype $ @@;
            do k = 1 to 3;
                input weeks @@;
                  do l = 1 to 2;
                      input gender $ @@;
                        do h = 1 to 2;
                            input liverse @@;
                            output;
                          end;
                    end;
              end;
        end;
  end;
 drop h; drop i; drop j; drop k; drop l;
run;
data b; set selenium;
if diettype = 'a' then diet = 1;
  else diet = 0;
if gender = 'f' then sex = 1;
  else sex = 0;
run;
proc glm;
  model liverse=diet sex dietse weeks
      diet*sex diet*dietse diet*weeks
      sex*dietse sex*weeks dietse*weeks;
run;
```

Figure 12.20 shows the GLM program that produced Rao's results in Example 12.6. The output from this program (with the Type I analysis edited out) is given in Figure 12.21. Note that if the Type III sums of squares and F-tests were not required, the same analysis of variance and parameter estimates could have been obtained by substituting REG for GLM in the program in Figure 12.20, since both of the categorical variables have

Figure 12.21. Analysis of selenium data

```
                    General Linear Models Procedure

Dependent Variable: LIVERSE

Source                 DF    Sum of Squares    F Value      Pr > F

Model                  10      37.53366979      47.73       0.0001

Error                  61       4.79705799

Corrected Total        71      42.33072778

             R-Square              C.V.            LIVERSE Mean

             0.886677           13.69244            2.04805556


Source                 DF      Type III SS    F Value      Pr > F

DIET                   1        0.19569670       2.49       0.1199
SEX                    1        0.73440247       9.34       0.0033
DIETSE                 1        0.42489218       5.40       0.0234
WEEKS                  1        2.88937245      36.74       0.0001
DIET*SEX               1        0.06480000       0.82       0.3676
DIET*DIETSE            1        0.13230000       1.68       0.1995
DIET*WEEKS             1        0.52920000       6.73       0.0119
SEX*DIETSE             1        0.37807500       4.81       0.0322
SEX*WEEKS              1        1.80187500      22.91       0.0001
DIETSE*WEEKS           1        0.81600312      10.38       0.0020


                                 T for H0:   Pr > |T|   Std Error of
Parameter          Estimate    Parameter=0                Estimate

INTERCEPT        0.173993056         0.73     0.4697     0.23917641
DIET             0.351666667         1.58     0.1199     0.22292666
SEX              0.681250000         3.06     0.0033     0.22292666
DIETSE           2.375520833         2.32     0.0234     1.02197917
WEEKS            0.368020833         6.06     0.0001     0.06071461
DIET*SEX         0.120000000         0.91     0.3676     0.13219539
DIET*DIETSE     -1.050000000        -1.30     0.1995     0.80952813
DIET*WEEKS      -0.105000000        -2.59     0.0119     0.04047641
SEX*DIETSE      -1.775000000        -2.19     0.0322     0.80952813
SEX*WEEKS       -0.193750000        -4.79     0.0001     0.04047641
DIETSE*WEEKS     0.798437500         3.22     0.0020     0.24786636
```

been changed to dummy variables in this program. Equivalent output could have been obtained by leaving both diet and sex as categorical variables, including a CLASS statement after PROC GLM; and including /SOLUTION in the MODEL statement.

In case the multiple DO statements in Figure 12.20 look overwhelming, refer to the SELENIUM.DAT data file, as shown in Figure 12.22.

Figure 12.22. The SELENIUM.DAT data file

```
.1 a 1 f 1.21 1.18 m 1.13 1.16 3 f 1.69 1.70 m 1.62 1.58
5 f 2.33 2.40 m 2.54 2.63 b 1 f 1.29 1.36 m 1.15 1.03
3 f 1.69 1.53 m 1.78 1.60 5 f 1.94 2.03 m 2.83 2.94
.2 a 1 f 1.59 1.56 m 1.21 1.29 3 f 2.00 1.93 m 1.73 1.82
5 f 2.11 2.08 m 2.56 2.67 b 1 f 1.25 1.46 m 1.06 1.14
3 f 1.83 1.90 m 1.77 1.61 5 f 2.87 3.01 m 2.98 3.23
.3 a 1 f 1.65 1.59 m 1.81 1.70 3 f 1.97 2.03 m 1.89 1.93
5 f 2.85 2.76 m 3.95 3.84 b 1 f 1.71 1.61 m 1.68 1.60
3 f 1.70 1.82 m 2.53 2.65 5 f 3.13 3.46 m 4.35 4.25
```

12.5 Analysis of Covariance

Example 12.7. In Example 12.3, the primary objective was to compare the responses (posttest scores) to three treatments designed to enhance children's mental capacities. Suspecting that the mental capacities of the children before they received the treatment may influence their posttest scores the investigators wanted to compare the treatment effects on the basis of the posttest scores of children with similar pretreatment mental capacities. Since the pretest score is an indicator of pretreatment mental capacity, the investigators decided to compare the treatments after adjusting the posttest scores to account for the differences between the pretest score.

Example 12.9. ...The analysis of posttest scores was based on the assumption that the relationship between the expected response and the level of the confounding factor can be represented by straight lines with a common slope. As we know, the assumption of parallel lines implies that there is no interaction between the treatments and the pretest scores, and so the general linear model in Example 12.3 is also an analysis of covariance model.

Rao's analysis of the pretest-posttest data as an analysis of covariance can be reproduced by Figure 12.12, in which a dummy coding was adopted to designate the three treatments. An equivalent analysis can be obtained using a CLASS statement in GLM with treatments as a categorical variable. Additionally, GLM will compute adjusted treatment means, which it calls "least squares means", if the LSMEANS statement is used, as is shown in Figure 12.23. For purposes of comparison, both the adjusted and unadjusted treatment means are requested in Figure 12.23.

Figure 12.23. Program to perform analysis of covariance and compute adjusted treatment means

```
options ls=70 ps=54;
/*Pretest-Posttest in 3 Treatments*/
data a;
  input tmt @@;
    do i = 1 to 10;
       input pre post @@;
       output;
    end;
drop i;
lines;
1 24 45 28 50 38 59 42 60 24 47 39 66 45 76
             19 50 19 39 22 36
2 23 28 33 39 31 36 34 39 18 22 24 28 41 49
             34 39 30 33 39 43
3 27 34 27 31 44 55 38 43 32 44 26 28 24 33
             13 13 36 39 52 58
;
run;
proc glm;
   class tmt;
   model post = tmt pre/solution;
   means tmt;
   lsmeans tmt/stderr pdiff;
run;
```

The program in Figure 12.23 produces the same analysis of variance table, Type I and Type III analyses, and parameter estimates as in Figure 12.12, so that portion of the output is not repeated in Figure 12.24. Additionally, the Class Levels Information is omitted from Figure 12.24. The differences in the unadjusted posttest means reflect, to some extent, the

Figure 12.24. **Analysis of covariance: adjusted and unadjusted treatment means**

```
                    General Linear Models Procedure
Level of     ----------POST----------   -----------PRE-----------
TMT       N     Mean         SD            Mean          SD

1        10   52.8000000   12.4078828   30.0000000    9.9777530
2        10   35.6000000    8.0027773   30.7000000    7.2118729
3        10   37.8000000   13.1892212   31.9000000   11.0900356

                        Least Squares Means

        TMT          POST        Std Err     Pr > |T|    LSMEAN
                    LSMEAN        LSMEAN    H0:LSMEAN=0   Number

         1       53.7780602    1.2116984     0.0001        1
         2       35.7880885    1.2099290     0.0001        2
         3       36.6338513    1.2124723     0.0001        3

            Pr > |T| H0: LSMEAN(i)=LSMEAN(j)

              i/j    1       2       3
               1     .     0.0001  0.0001
               2   0.0001    .     0.6257
               3   0.0001  0.6257    .
```

NOTE: To ensure overall protection level, only probabilities
      associated with pre-planned comparisons should be used.

fact that the three groups had differing pretest means. The adjusted treatment means reflect what the posttest means would have been if it had been possible to fix it so that pretest scores were

$$(30.0 + 30.7 + 32.8)/3$$

$$= 31.17$$

in each of the three groups. Thus, the means in the first two groups are adjusted upward, since these two groups started out below average, but the mean in the last group is adjusted downward to take away its pretest advantage. That is, for Group i,

$$\overline{Y}_{i(adj)} = \overline{Y}_i - b_1(\overline{x}_i - \overline{x}),$$

where $\overline{Y}_i$ is the unadjusted posttest score for treatment i, $\overline{x}_i$ is the pretest score in treatment i, and $\overline{x}$ is the mean for all pretest scores, averaged over treatments (30.8667, in this example).

Recall from Figure 12.12 that the slope of the line relating posttest score to pretest score was $b_1 = 1.1285$ for each of the three treatments. Then

$$\overline{Y}_{1(adj)} = \overline{Y}_1 - b_1(\overline{x}_1 - \overline{x})$$
$$= 52.8 - 1.1285(30.0 - 30.8667)$$
$$= 52.8 - 1.1285(-0.8667)$$
$$= 52.8 + 0.9780$$
$$= 53.8,$$

and similarly for the other two groups.

Equivalently, the adjusted treatment means can be obtained by substituting the overall mean pretest score, 30.8667, into the equation relating posttest score to pretest score for each treatment. For example, Figure 12.12 gives the equation for Treatment 1 as

$$Post = 18.9441 + 1.1285(Pre),$$

so that

$$Post = 18.9441 + 1.1285(30.8667)$$

$$= 53.8.$$

Referring again to the program in Figure 12.23, note the two options in the LSMEANS command. STDERR results in the standard errors of the adjusted means being printed, and a test of the null hypothesis that the (population) adjusted treatment mean is zero in each group. These are probably not of interest in this example. PDIFF stands for the p-value for testing whether two adjusted treatment means differ significantly from each other. The output in Figure 12.24 shows a matrix of such p-values. The diagonal elements are blank, and the matrix is symmetric. We see that the adjusted mean for Treatment 1 differs significantly from the other two adjusted means, but that the adjusted means of Treatments 2 and 3 do not differ significantly from each other. The note at the bottom of the printout reminds us that the probability of a Type I error increases as more tests are made, and that p-values do not apply to hypotheses suggested by the data.

13

Analysis of Multifactor Studies: Factorial and Nested Experiments in Completely Randomized Designs

13.1 Introduction

The ANOVA and GLM procedures are used to analyze multifactor factorial and nested designs. In addition to testing main effects in the factorial designs, sub-effects of the interaction effects are tested or estimated.

13.2 Analyzing a 2 x 2 Factorial Experiment

Example 13.6. The percentage of water content in the tissues of snails grown under six different experimental conditions was measured. The six conditions were obtained by combining three levels of humidity (factor B) with two levels of temperature (factor A). There were $n = 4$ replications of each treatment combination... The tissue water content data can be regarded as the results from a 3x2 factorial experiment.

Example 13.8. The following are the treatment totals for the tissue water content data (with the data for 100% humidity deleted) of Example 13.6.

Since this is a balanced design, either the ANOVA or the GLM procedure can be used to perform the analysis of variance. Note, however, that ANOVA can be used only for balanced multifactor studies: unbalanced multifactor studies must be analyzed using GLM.

The program in Figure 13.1 will produce the analysis of variance, a table of row, column, and cell means, and two different interaction plots using the cell means. Note that the data at 100% humidity have been excluded from the analysis by the specification

```
where humidity ne 100;.
```

Figure 13.1. Program to analyze a factorial experiment

```
options ls=70 ps=54;
/*2 x 3 factorial*/
data;
input temp @;
do i = 1 to 3;
input humidity @;
do j = 1 to 4;
input water @;
output;
end;
end;
lines;
20 45 76 64 79 71 75 72 82 86 86 100 100 96
92 100
30 45 72 72 64 70 75 72 75 82 84 100 100 94
98   99
;
run;
proc glm;
class temp humidity;
model water = temp humidity temp*humidity;
means temp humidity temp*humidity;
where humidity ne 100;
output out=avgs p=ybar;
run;
proc plot data=avgs;
plot ybar*humidity=temp;
plot ybar*temp=humidity;
run;
```

The program in Figure 13.1 is quite similar to the one in Figure 8.1 for analyzing data from a one-way study. Here, however, the names of both factors are given in the CLASS statement, and both factors *and their interaction*

```
temp*humidity
```

are defined in the model statement. Predicted values at each combination of levels of the two factors, which are simply cell means, are outputted to a data set named "avgs" so that they may be used to construct interaction plots.

Class Levels Information is given in Figure 13.2, and the analysis of variance is found in Figure 13.3.

Figure 13.2. Class levels information for a 2 × 2 factorial

```
                    General Linear Models Procedure
Class Level Information
                    Class     Levels     Values

                    TEMP        2        20 30

                    HUMIDITY    2        45 75


Number of observations in data set = 16
```

In Figure 13.3, we see a significant model F-value, and that variation in temperature and/or humidity and/or their interaction explain approximately 50% of the variation in water content. This Model Sum of Squares with its 3 degrees of freedom is decomposed into individual-degree-of-freedom sums of squares for each of the two main effects and the interaction. Since the experiment is balanced, both the Type I and the Type III decompositions are identical (and therefore redundant).   There is no significant interaction between temperature and humidity, and neither do there appear to be any differences in water content according to temperature. Almost all of the model variation is due to humidity differences.

Figure 13.3. Analysis of variance for a 2 × 2 factorial

```
                    General Linear Models Procedure
Dependent Variable: WATER
Source                DF     Sum of Squares    F Value    Pr > F

Model                  3      354.18750000      3.54      0.0480

Error                 12      399.75000000

Corrected Total       15      753.93750000


              R-Square              C.V.            WATER Mean

              0.469784            7.650967          75.4375000



Source                DF        Type I SS       F Value    Pr > F

TEMP                   1       39.06250000       1.17      0.3001
HUMIDITY               1      315.06250000       9.46      0.0096
TEMP*HUMIDITY          1        0.06250000       0.00      0.9662

Source                DF       Type III SS      F Value    Pr > F

TEMP                   1       39.06250000       1.17      0.3001
HUMIDITY               1      315.06250000       9.46      0.0096
TEMP*HUMIDITY          1        0.06250000       0.00      0.9662
```

Figure 13.4. Means for factor levels and treatment combinations

```
                 General Linear Models Procedure
         Level of      ------------WATER------------
         TEMP      N        Mean               SD

         20        8     77.0000000        7.76438757
         30        8     73.8750000        6.46832944

         Level of      ------------WATER------------
         HUMIDITY  N        Mean               SD

         45        8     71.0000000        5.20988072
         75        8     79.8750000        5.96268156

  Level of   Level of      ------------WATER------------
  TEMP       HUMIDITY  N        Mean               SD

  20         45        4     72.5000000        6.55743852
  20         75        4     81.5000000        6.60807587
  30         45        4     69.5000000        3.78593890
  30         75        4     78.2500000        5.67890835
```

Figure 13.5. Cell means plotted against humidity

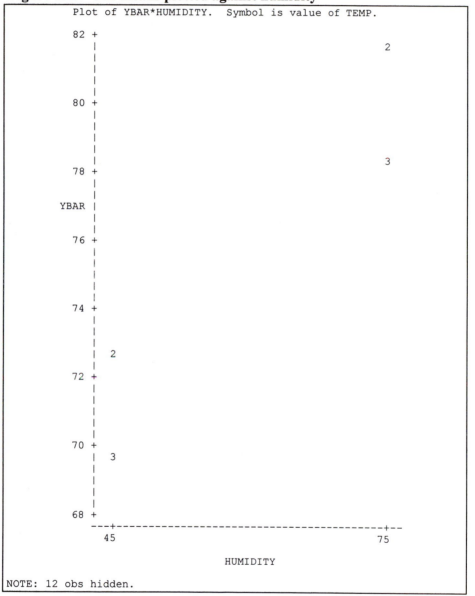

```
         Plot of YBAR*HUMIDITY.   Symbol is value of TEMP.

    82 +
       |                                                    2
       |
       |
       |
       |
    80 +
       |
       |
       |
       |
    78 +                                                    3
       |
 YBAR  |
       |
       |
    76 +
       |
       |
       |
       |
    74 +
       |
       |
       |
       |  2
    72 +
       |
       |
       |
       |
    70 +
       |  3
       |
       |
       |
       |
    68 +
       ---+-------------------------------------------+--
          45                                          75
                          HUMIDITY
```

NOTE: 12 obs hidden.

Figure 13.6. Cell means plotted against temperature

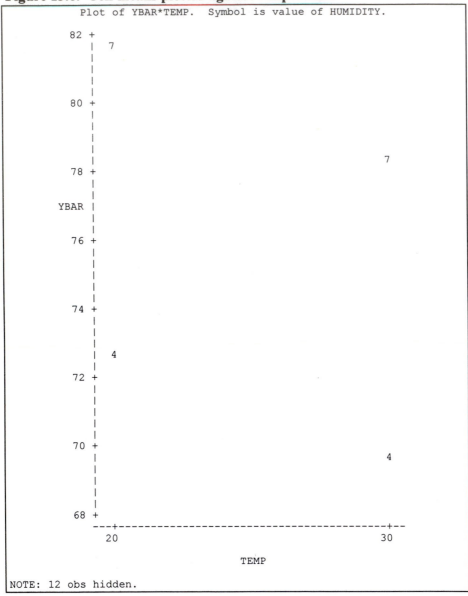

```
          Plot of YBAR*TEMP.   Symbol is value of HUMIDITY.

     82 +
        |    7
        |
        |
        |
        |
     80 +
        |
        |
        |
        |
        |                                                        7
     78 +
        |
 YBAR   |
        |
        |
     76 +
        |
        |
        |
        |
     74 +
        |
        |
        |    4
        |
     72 +
        |
        |
        |
        |
     70 +
        |                                                        4
        |
        |
        |
     68 +
        ---+-----------------------------------------------+--
           20                                               30

                                TEMP

NOTE: 12 obs hidden.
```

Figure 13.4 shows mean water content for both levels of temperature, both levels of humidity, and at all four combinations of levels of temperature and humidity. The cell means are graphed in Figures 13.5 and 13.6. Both graphs show parallelism, indicating lack of interaction, and both show large average differences between the two humidity levels but small differences due to temperature.

Since there is no significant interaction between temperature and humidity, mean responses at the different levels of these factors may be compared by writing contrasts as was done for single-factor studies. However, in this example, temperature effects are nonsignificant. Further, no additional information would be obtained by writing a CONTRAST statement to compare the two levels of humidity, since the contrast sum of squares will be identical to the Type I and Type III sums of squares for humidity in Figure 13.3. However, it might be of interest to use the ESTIMATE statement to estimate the difference in the mean response at the two humidity levels. The following code, included under GLM, will produce the estimate, its standard error, and associated *t*- and *p*-values:

```
estimate 'HivsLo' humidity -1 1;.
```

13.3 Analyzing an *A* x *B* Factorial Experiment

Example 13.9. In Example 13.6, we described a factorial experiment with two factors: *A*, the temperature, at $a = 2$ levels (20°C and 30°C); and *B,* the humidity, at $b = 3$ levels (45%. 75% and 100%). There were $n = 4$ replications per treatment, resulting in $N = abn = 24$ responses. The objective of the experiment was to study the effects of temperature and humidity on the tissue water content of snails.

The SAS program shown in Figure 13.1, with the omission of the line

```
where humidity ne 100;
```

will analyze the data for this 2×3 factorial experiment. Class Levels Information and the analysis of variance are shown in Figure 13.7; factor level means and cell means are given in Figure 13.8; and interaction plots are shown in Figures 13.9 and 13.10.

Figure 13.7. Analysis of a 2×3 factorial experiment.

```
                   General Linear Models Procedure
                       Class Level Information
                    Class    Levels    Values

                    TEMP        2      20 30

                    HUMIDITY    3      45 75 100


Number of observations in data set = 24

                   General Linear Models Procedure
Dependent Variable: WATER
Source                   DF    Sum of Squares    F Value    Pr > F

Model                     5    2922.00000000      22.65     0.0001

Error                    18     464.50000000

Corrected Total          23    3386.50000000

              R-Square              C.V.              WATER Mean

              0.862838            6.138872            82.7500000


Source                   DF       Type I SS      F Value    Pr > F

TEMP                      1      20.16666667       0.78      0.3883
HUMIDITY                  2    2881.75000000      55.84      0.0001
TEMP*HUMIDITY             2      20.08333333       0.39      0.6832

Source                   DF      Type III SS     F Value    Pr > F

TEMP                      1      20.16666667       0.78      0.3883
HUMIDITY                  2    2881.75000000      55.84      0.0001
TEMP*HUMIDITY             2      20.08333333       0.39      0.6832
```

Figure 13.8. Cell and factor level means in the 2 × 3 factorial experiment

```
               General Linear Models Procedure
        Level of              ------------WATER------------
        TEMP           N         Mean                SD

         20            12      83.6666667         11.8039541
         30            12      81.8333333         12.9111181

        Level of              ------------WATER------------
        HUMIDITY       N         Mean                SD

         45             8      71.0000000          5.20988072
         75             8      79.8750000          5.96268156
        100             8      97.3750000          3.06768875

Level of   Level of            ------------WATER------------
TEMP       HUMIDITY      N         Mean                SD

 20          45          4       72.5000000         6.55743852
 20          75          4       81.5000000         6.60807587
 20         100          4       97.0000000         3.82970843
 30          45          4       69.5000000         3.78593890
 30          75          4       78.2500000         5.67890835
 30         100          4       97.7500000         2.62995564
```

Since Figure 13.7 indicates that there is no interaction between levels of temperature and levels of humidity, comparison of factor levels for main effects can be done just as for single-factor studies. The small F-value for TEMP indicates no statistically significant difference in water content according to temperature levels, but the large F-value for HUMIDITY indicates differences in the mean response among some levels of humidity. In order to ascertain which humidity levels produce results that are significantly different from others, and how, either (1) *a priori* contrasts among the three humidity levels may be written, (2) all pairwise comparisons among humidity level means may be made, or (3) some differences suggested by the data can be explored for significance. For example, to test whether the mean responses at 45% and 75% differ significantly, write

```
contrast '75vs45' humidity -1 1 0;
```

Figure 13.9. Cell means plotted against temperature levels

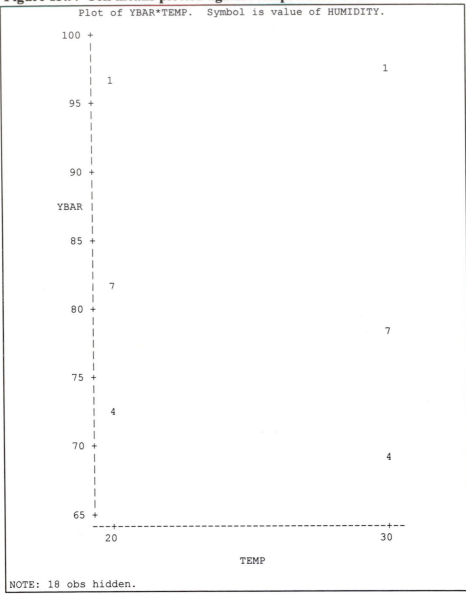

```
              Plot of YBAR*TEMP.   Symbol is value of HUMIDITY.

        100 +
            |
            |
            |                                                          1
            |    1
            |
         95 +
            |
            |
            |
            |
         90 +
            |
   YBAR     |
            |
            |
         85 +
            |
            |
            |
            |    7
            |
         80 +
            |
            |                                                      7
            |
            |
            |
         75 +
            |
            |
            |    4
            |
            |
         70 +
            |                                                      4
            |
            |
            |
            |
         65 +
            ---+-------------------------------------------------+--
               20                                                30

                                      TEMP
```

NOTE: 18 obs hidden.

Figure 13.10. Cell means plotted against humidity levels

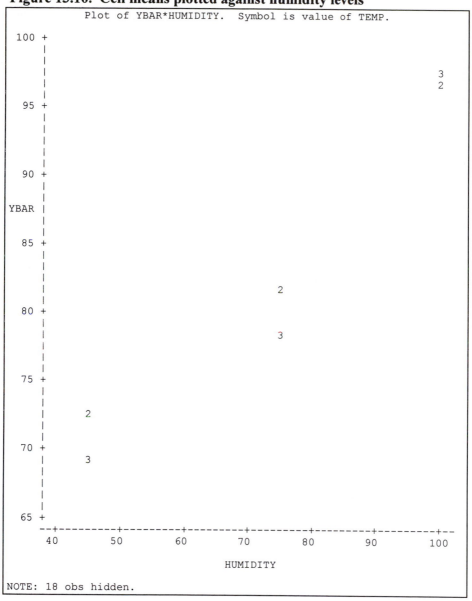

NOTE: 18 obs hidden.

in the GLM procedure. To estimate how large the difference is in the mean response at 100% and at 75% humidity, include

```
estimate '100vs75' humidity 0 -1 1;.
```

To compare means at each of the three humidity levels to both of the others, use the Fisher's least significant difference method For example, write

```
means humidity/lsd;.
```

13.4 Analysis of Factorial Experiments with Interaction

Example 13.11.   In Example 9.14, we described a 2 × 2 factorial experiment to study the effect of water temperature and water movement on the weight gain of fish. The treatments were combinations of two levels of water temperature (cold and warm) and two levels of water movement (still and flowing).

The weight gain data in Example 9.14 did not include data for a third level of temperature investigated in the original study. The actual study was a 3 × 2 factorial, in which there were three levels of water temperature (cold, lukewarm, warm) and two levels of water movement (still, flowing)

Interaction

As was the case in Chapter 9, we are given summary statistics, but not the actual data for this experiment. However, it is possible to "re-construct" a "data" set that has the same treatment means and pooled within-treatment variance as the actual data. The computer can thus be "tricked" into producing the desired analysis. The program in Figure 13.11 constructs a "pseudo" data set with the required properties. The only change from the program in Figure 9.5 for reconstructing data for a one-way analysis is that

Figure 13.11. Program to reconstruct and analyze data from a two-way experiment

```
options ls=70 ps=54;
/*Interaction*/
data a;
   input temp$ move$ n ybar sd;
   delta = sqrt((n - 1)/2) * sd;
   weight = n - 2; y = ybar; output;
   weight = 1; y = ybar - delta; output;
   y = ybar + delta; output;
lines;
cold still 3 1.55 0.1581138
cold flowing 3 1.08 0.1581138
luke still 3 1.87 0.1581138
luke flowing 3 1.76 0.1581138
warm still 3 1.59 0.1581138
warm flowing 3 2.01 0.1581138
;
proc print; run;
proc glm;
   class temp move;
   model y = temp move temp*move;
   output out=avgs p = ybar;
   means temp move temp*move;
run;
proc plot;
   plot ybar*temp = move;
   plot ybar*move=temp;
run;
```

the variable GROUP is replaced in the CLASS statement by the two factors *A* and *B* and by

A B A*B

in the MODEL statement. For your information, the "pseudo" data set is displayed in Figure 13.12. We see in Figure 13.13 that when the "data" set is constructed in such a manner, it has the same properties as the actual data set, apart from some small rounding error.

Figure 13.12. Pseudo data set for two-factor experiment

OBS	TEMP	MOVE	N	YBAR	SD	DELTA	WEIGHT	Y
1	cold	still	3	1.55	0.15811	0.15811	1	1.55000
2	cold	still	3	1.55	0.15811	0.15811	1	1.39189
3	cold	still	3	1.55	0.15811	0.15811	1	1.70811
4	cold	flowing	3	1.08	0.15811	0.15811	1	1.08000
5	cold	flowing	3	1.08	0.15811	0.15811	1	0.92189
6	cold	flowing	3	1.08	0.15811	0.15811	1	1.23811
7	luke	still	3	1.87	0.15811	0.15811	1	1.87000
8	luke	still	3	1.87	0.15811	0.15811	1	1.71189
9	luke	still	3	1.87	0.15811	0.15811	1	2.02811
10	luke	flowing	3	1.76	0.15811	0.15811	1	1.76000
11	luke	flowing	3	1.76	0.15811	0.15811	1	1.60189
12	luke	flowing	3	1.76	0.15811	0.15811	1	1.91811
13	warm	still	3	1.59	0.15811	0.15811	1	1.59000
14	warm	still	3	1.59	0.15811	0.15811	1	1.43189
15	warm	still	3	1.59	0.15811	0.15811	1	1.74811
16	warm	flowing	3	2.01	0.15811	0.15811	1	2.01000
17	warm	flowing	3	2.01	0.15811	0.15811	1	1.85189
18	warm	flowing	3	2.01	0.15811	0.15811	1	2.16811

Figure 13.13 indicates that there is a highly significant interaction between temperature and movement. That is, the way temperature affects weight gain *depends on* whether the water is still or is moving, and vise versa. How the effect of temperature varies according to what the water is doing may be seen in the factor level means and cell means in Figure 13.14, but are more easily seen in the graphs in Figures 13.15 and 13.16.

Figure 13.15 shows that as long as the water temperature is cold or lukewarm, weight gains are greater for fish grown in still water than for those grown in flowing water. However, the opposite is true for fish grown in warm water. Weight gains for these fish are greater if the water is flowing instead of still. Equivalently, Figure 13.16 indicates that for fish grown in flowing water, weight gains increase with water temperature, but for those grown in still water, weight gains are greatest under lukewarm water.

That the interaction is *qualitative*, rather than quantitative, is what makes it appear that there is no effect of water movement on weight gains. There *is* an effect of water movement: it is just *different* at different water

Figure 13.13. Analysis of variance from analysis of reconstructed two-way data

```
                    General Linear Models Procedure
                        Class Level Information

               Class     Levels     Values

               TEMP          3      cold luke warm

               MOVE          2      flowing still


              Number of observations in data set = 18
```

Dependent Variable: Y

Source	DF	Sum of Squares	F Value	Pr > F
Model	5	1.58500000	12.68	0.0002
Error	12	0.29999969		
Corrected Total	17	1.88499969		

R-Square	C.V.	Y Mean
0.840849	9.621529	1.64333333

Source	DF	Type I SS	F Value	Pr > F
TEMP	2	0.97090000	19.42	0.0002
MOVE	1	0.01280000	0.51	0.4880
TEMP*MOVE	2	0.60130000	12.03	0.0014

Source	DF	Type III SS	F Value	Pr > F
TEMP	2	0.97090000	19.42	0.0002
MOVE	1	0.01280000	0.51	0.4880
TEMP*MOVE	2	0.60130000	12.03	0.0014

temperatures. Flowing water increases weight gain for fish grown in warm water but decreases weight gain for those grown in cold or lukewarm water. Therefore, if you average the effects of water movement across all levels of temperature, the movement effects cancel each other. This opposite effect at different levels is what makes it appear that there is no effect of movement.

Figure 13.14. Factor level means and cell means

```
              General Linear Models Procedure

         Level of              --------------Y--------------
         TEMP         N           Mean              SD

         cold         6        1.31500000        0.29371752
         luke         6        1.81500000        0.15372046
         warm         6        1.80000000        0.27003700

         Level of              --------------Y--------------
         MOVE         N           Mean              SD

         flowing      9        1.61666667        0.43871971
         still        9        1.67000000        0.20383812

Level of    Level of            --------------Y--------------
TEMP        MOVE         N          Mean              SD

cold        flowing      3       1.08000000        0.15811380
cold        still        3       1.55000000        0.15811380
luke        flowing      3       1.76000000        0.15811380
luke        still        3       1.87000000        0.15811380
warm        flowing      3       2.01000000        0.15811380
warm        still        3       1.59000000        0.15811380
```

Since for a qualitative interaction, such as the one illustrated by these data, main effects of one factor tend to cancel each other when averaged across the levels of the other factor, it is inappropriate to analyze main effects as was done for single-factor studies or for multifactor studies in which no interaction is present. Instead of contrasting temperature levels averaging across levels of movement, it is necessary to compare effects of the various temperatures *within* each of the levels of movement. Similarly, instead of estimating the difference in weight gains under still and flowing water, averaging across temperatures, we need to compare the effects of still and flowing water *within* each temperature level. Even if the interaction had been quantitative, instead of qualitative, the same argument can be made: in the presence of interaction, main effects of one factor must be analyzed *holding constant* the level of the other factor or factors.

Figure 13.15. Plot of cell means against movement

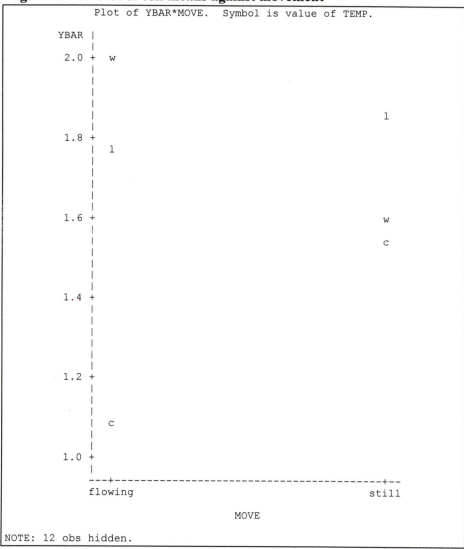

```
            Plot of YBAR*MOVE.   Symbol is value of TEMP.
     YBAR |
          |
      2.0 +  w
          |
          |
          |
          |
          |                                            l
          |
      1.8 +
          |  l
          |
          |
          |
          |
          |
      1.6 +                                            w
          |
          |                                            c
          |
          |
          |
          |
      1.4 +
          |
          |
          |
          |
          |
          |
      1.2 +
          |
          |
          |  c
          |
          |
      1.0 +
          |
          ---+----------------------------------------+--
          flowing                                   still

                                MOVE

NOTE: 12 obs hidden.
```

Figure 13.16. Plot of cell means against temperature

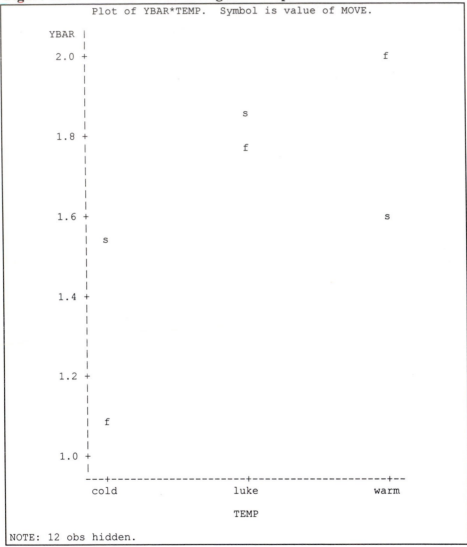

```
            Plot of YBAR*TEMP.   Symbol is value of MOVE.

   YBAR |
        |
    2.0 +                                                    f
        |
        |
        |
        |                              s
    1.8 +
        |                              f
        |
        |
        |
        |
    1.6 +                                                    s
        |
        |    s
        |
        |
        |
    1.4 +
        |
        |
        |
        |
        |
    1.2 +
        |
        |
        |    f
        |
        |
    1.0 +
        |
        ---+--------------------+--------------------+--
          cold                 luke                 warm

                              TEMP
NOTE: 12 obs hidden.
```

Contrasts and Estimates of Interaction Effects

Suppose we wish to investigate the nature of the interaction between water temperature and water movement by making the following comparisons:

1. Effects of still vs flowing water at cold temperature
2. Effects of still vs flowing water at lukewarm temperature
3. Effects of still vs flowing water at warm temperature
4. Differences among temperature levels in still water
5. Differences among temperature levels in flowing water

The first three comparisons involve simple contrasts between two movement means, holding temperature constant. Each of these comparisons will have one degree of freedom in the numerator of the *F*-test, and will require one line to write contrast constants in the CONTRAST or ESTIMATE statement in GLM. The last two comparisons above look for any differences at the three temperature levels, but holding water movement constant. These latter comparisons will be based on two degrees of freedom, as they would be in two separate single-factor ANOVAs, and will require two lines of contrast constants in the CONTRAST or ESTIMATE command.

First, recall that when factor levels are entered as character variables, SAS treats them in alphabetical order. Thus, the levels of temperature are 1 = cold, 2 = lukewarm, and 3 = warm, and the levels of water movement are 1 = flowing and 2 = still. There are six combinations of temperature and movement levels. *If in the MODEL statement, TEMP is listed first,* then the six treatments correspond to

$$11, 12, 21, 22, 31, 32$$

or

cold flowing, cold still, lukewarm flowing,

lukewarm still, warm flowing, warm still.

Had the model statement listed MOVE first, then SAS would have considered the six treatments in the order

$$11, 12, 13, 21, 22, 23$$

or

flowing cold, flowing lukewarm, flowing warm,
still cold, still lukewarm, still warm.

In order to write CONTRAST or ESTIMATE statements which address interaction effects, it is necessary to specify the levels of factor A, the levels of factor B, and the interaction effects involved in the comparison. It takes a bit of algebra to determine what these are. Recall the model for the two-factor factorial experiment:

$$E(y_{ij}) = \mu + \alpha_i + \beta_j + (\alpha\beta)_{ij}.$$

Assume that water temperature is factor A and will be listed first in the GLM MODEL statement. Then the expected responses at cold, flowing water and at cold still water are, respectively,

$$E(y_{11}) = \mu + \alpha_1 + \beta_1 + (\alpha\beta)_{11}$$

and

$$E(y_{12}) = \mu + \alpha_1 + \beta_2 + (\alpha\beta)_{12}.$$

Thus, the difference in expected responses for still and flowing cold water is

$$E(y_{11}) - E(y_{12}) = \beta_1 - \beta_2 + (\alpha\beta)_{11} - (\alpha\beta)_{12}.$$

We see that this difference does not involve any of the level of factor A, but that the coefficient on level 1 of factor B is +1, the coefficient on level 2 of factor *B* is −1, the (1,1) interaction term has coefficient +1, the (1,2) interaction term has coefficient −1, and all other interaction terms have coefficients zero. Thus, the contrast (or estimate) statement is written

```
contrast 'move in cold'
        temp 0 0 0 move 1 -1 temp*move 1 -1 0 0 0 0;.
```

The portion

```
                temp 0 0 0
```

could have been omitted, as could have the trailing zeroes in

```
        temp*move 1 -1 0 0 0 0.
```

A similar bit of algebra indicates that the differences in expected responses for still and flowing water at lukewarm and warm temperatures are, respectively,

$$E(y_{21}) - E(y_{22}) = \beta_1 - \beta_2 + (\alpha\beta)_{21} - (\alpha\beta)_{22}$$

and

$$E(y_{31}) - E(y_{32}) = \beta_1 - \beta_2 + (\alpha\beta)_{31} - (\alpha\beta)_{32} .$$

The contrasts to make these last two comparisons are thus written

```
contrast 'move in luke'
        temp 0 0 0 move 1 -1 temp*move 0 0 1 -1 0 0;
```

and

```
contrast 'move in warm'
        temp 0 0 0 move 1 -1 temp*move 0 0 0 0 1 -1;.
```

We see that all three comparisons involve the difference

$$\beta_1 - \beta_2 ,$$

which compares results under still and flowing water conditions. None of the comparisons involve comparing temperatures, however; instead temperatures are held constant, so α_1, α_2, or α_3 do not appear, except through their involvement in the interaction terms.

Figure 13.17 shows the output from these three CONTRAST commands. While water movement does not appear to affect weight gain for fish grown in lukewarm water, there are effects of water movement on fish grown in cold and in warm water. Replacing the word CONTRAST with the word ESTIMATE in the above commands produces estimates of the differences, as shown in Figure 13.18. We see that the mean weight gain under flowing water is less than under still water if the water is cold, but that the opposite is true for fish grown in warm water.

Figure 13.17. Contrasts between movement levels for each of the three temperature levels

Contrast	DF	Contrast SS	F Value	Pr > F
move in cold	1	0.33135000	13.25	0.0034
move in luke	1	0.01815000	0.73	0.4109
move in warm	1	0.26460000	10.58	0.0069

Figure 13.18. Estimates of the effects of movement within each temperature level

Parameter	Estimate	T for H0: Parameter=0	Pr > |T|	Std Error of Estimate
move in cold	-0.47000000	-3.64	0.0034	0.12909938
move in luke	-0.11000000	-0.85	0.4109	0.12909938
move in warm	0.42000000	3.25	0.0069	0.12909938

To test the null hypothesis that there are no differences due to temperature when water is flowing requires an F-test with two degrees of freedom in the numerator, since three temperatures are being compared. In order to compare three levels to each other, choose one of the levels as a reference and compare the other two levels to it. The CONTRAST command will require two sets of constants, separated by commas, one for each comparison. For example, to test

$$H_0 : \text{no temperature differences in flowing water}$$

the CONTRAST command could be written as

```
contrast 'temps in flow'
        temp 1 0 -1 temp*move 1 0 0 0 -1 0,
        temp 0 1 -1 temp*move 0 0 1 0 -1 0;
```

if using warm water as a reference. Equivalently, the same test could be obtained by writing either

```
contrast 'temps in flow'
        temp 1 -1 0 temp*move 1 0 -1 0 0 0,
        temp 0 -1 1 temp*move 0 0 -1 0 1 0;
```

or

```
contrast 'temps in flow'
        temp -1 1 0 temp*move -1 0 1 0 0 0,
        temp -1 0 1 temp*move -1 0 0 0 1 0;.
```

To test

$$H_0 : \text{no temperature differences in still water}$$

a CONTRAST command might be

```
contrast 'temps in still'
        temp 1 0 -1 temp*move 0 1 0 0 0 -1,
        temp 0 1 -1 temp*move 0 0 0 1 0 -1;.
```

Figure 13.19 shows the contrast sums of squares and F-tests generated. These tests show that there are differences in weight gain at the three different temperature levels for fish grown in still water and for fish grown in flowing water, although the differences are apparently much more pronounced for those fish grown in flowing water.

Figure 13.19. Temperature effects for still and flowing water separately

Contrast	DF	Contrast SS	F Value	Pr > F
temps in still	2	0.18240000	3.65	0.0578
temps in flow	2	1.38980000	27.80	0.0001

Inspection of the interaction plots in Figures 13.15 and 13.16 might suggest some (after-the-fact) hypotheses to test in order to investigate the temperature effects in greater depth. Since the plot of treatment means against temperature indicates an increasing trend in weight gain with temperature for fish grown in flowing water, one might want to compare mean weight gain at lukewarm flowing to that at cold flowing and mean gain at warm flowing to that at lukewarm flowing. These two contrasts would be written

```
        contrast 'lukeflow vs coldflow'
            temp -1 1 0 temp*move -1 0 1 0 0 0;
```
and

```
        contrast 'warmflow vs lukeflow'
            temp 0 -1 1 temp*move 0 0 -1 0 1 0;.
```

In addition, one might ask how the size of the increase from cold to lukewarm flowing water compares to the size of the increase from lukewarm to warm flowing water. Let the difference between warm and lukewarm flowing water be denoted by D_1, and let the difference at lukewarm and cold temperatures of flowing water be denoted by D_2. Referring back to the model for the two-way factorial,

$$D_1 = E(y_{31}) - E(y_{21}) = \alpha_3 - \alpha_2 + (\alpha\beta)_{31} - (\alpha\beta)_{21}$$
and

$$D_2 = \alpha_2 - \alpha_1 + (\alpha\beta)_{21} - (\alpha\beta)_{11}$$

so that the difference is

$$D_1 - D_2 = \alpha_3 - 2\alpha_2 + \alpha_1 + (\alpha\beta)_{31} - 2(\alpha\beta)_{21} + (\alpha\beta)_{11}.$$

Writing

```
estimate 'd1 vs d2'
        temp 1 -2 1 temp*move 1 0 -2 0 1 0;
```

produces the estimate in Figure 13.20. We see that the largest increase in fish weight in flowing water comes from increasing water temperature from cold to lukewarm.

Figure 13.20. Comparing temperature effects for flowing water condition

Parameter	Estimate	T for H0: Parameter=0	Pr > |T|	Std Error of Estimate
d1	0.25000000	1.94	0.0767	0.12909938
d2	0.68000000	5.27	0.0002	0.12909938
d1 vs d2	-0.43000000	-1.92	0.0785	0.22360668

Computing Sums of Squares for Main Effects and Interaction

At this point, we will illustrate how to compute the sums of squares for the two main effects and the interaction in the analysis of variance table in Figure 13.13. This will be a useful technique to know in the next section, in which a three-factor study is considered.

Consider SS[MOVE] = 0.0128, with 1 degree of freedom. This sum of squares compares the average of cell means for still water conditions to the average of cell means under flowing water conditions:

$$\text{SS[MOVE]} = (\mu_{12} + \mu_{22} + \mu_{32}) - (\mu_{11} + \mu_{21} + \mu_{31}).$$

A bit of algebra shows that this contrast is

$$3\beta_2 - 3\beta_1 + (\alpha\beta)_{12} + (\alpha\beta)_{22} + (\alpha\beta)_{32} - (\alpha\beta)_{11} - (\alpha\beta)_{21} - (\alpha\beta)_{31}.$$

Thus, the following contrast will produce the sum of squares for the MOVE main effect:

```
contrast 'move' move -3 3 temp*move -1 1 -1 1 -1 1;.
```

The sum of squares for TEMP has two degrees of freedom and will require comparison of two of the levels of movement to a third; for example,

$$\mu_{11} + \mu_{12} - (\mu_{31} - \mu_{32})$$

and

$$\mu_{21} + \mu_{22} - (\mu_{31} - \mu_{32}).$$

Again, algebraic manipulation can be used to express these contrasts as

$$2\alpha_1 - 2\alpha_3 + (\alpha\beta)_{11} + (\alpha\beta)_{12} - (\alpha\beta)_{31} - (\alpha\beta)_{32}$$

and

$$2\alpha_2 - 2\alpha_3 + (\alpha\beta)_{21} + (\alpha\beta)_{22} - (\alpha\beta)_{31} - (\alpha\beta)_{32}.$$

A contrast to produce SS[TEMP] is thus

```
contrast 'temp' temp 2 0 -2 temp*move 1 1 0 0 -1 -1,
                temp 0 2 -2 temp*move 0 0 1 1 -1 -1;.
```

Finally, the interaction sum of squares compares the difference between still and flowing water at each of two temperatures with the difference at the third temperature:

$$\mu_{11} - \mu_{12} - (\mu_{31} - \mu_{32})$$

and

$$\mu_{21} - \mu_{22} - (\mu_{31} - \mu_{32}),$$

which can be written as

$$(\alpha\beta)_{11} - (\alpha\beta)_{12} - (\alpha\beta)_{31} + (\alpha\beta)_{32}$$

and

$$(\alpha\beta)_{21} - (\alpha\beta)_{22} - (\alpha\beta)_{31} + (\alpha\beta)_{32} .$$

The SS[TEMP*MOVE] can be produced by

```
contrast 'interact' temp*move 1 -1 0 0 -1 1,
                     temp*move 0 0 1 -1 -1 1;.
```

13.5 Factorial Experiments with More than Two Factors

Example 13.14. In Example 13.11, we presented the weight gain data for fish grown under six experimental conditions obtained by combining three levels of water temperature (cold, lukewarm, warm) and two levels of water flow (still, flowing). The data were analyzed as the results of a 2 × 3 factorial experiment.

For the data in Example 13.11, the aquarium light condition was maintained at a low level (16 hrs/day). [Rao gives a] summary of the treatment totals [which] includes the data from Example 13.11 along with data for six new experimental conditions in which the aquarium light condition was maintained at a high level (24 hrs/day).

The SAS program to construct a "pseudo" data set for use in illustrating the analysis, to perform an analysis of variance using the GLM procedure, and to obtain interaction plots is shown in Figure 13.21. The MODEL statement in the GLM procedure specifies all three two-factor interactions and the three-factor interaction. Since it becomes tedious to list all possible interaction terms in multifactor factorial designs, a shortcut way to indicate to SAS that all factors are crossed with all others is to use the "bar operator":

```
model y = light|temp|move;.
```

Figure 13.21. Program for analyzing a three-factor factorial experiment

```
options ls=70 ps=54;
/*Interaction*/
data a;
  input light$ temp$ move$ n ybar sd;
  delta = sqrt((n - 1)/2) * sd;
  weight = n - 2; y = ybar; output;
  weight = 1; y = ybar - delta; output;
  y = ybar + delta; output;
lines;
low cold still 3 1.55 0.1516575
low cold flowing 3 1.08 0.1516575
low luke still 3 1.87 0.1516575
low luke flowing 3 1.76 0.1516575
low warm still 3 1.59 0.1516575
low warm flowing 3 2.01 0.1516575
high cold still 3 1.85 0.1516575
high cold flowing 3 1.36 0.1516575
high luke still 3 2.03 0.1516575
high luke flowing 3 1.67 0.1516575
high warm still 3 2.14 0.1516575
high warm flowing 3 1.85 0.1516575
;
proc print; run;
proc glm;
  class light temp move;
  model y = light temp move light*temp
        light*move temp*move light*temp*move;
  output out=avgs p = ybar;
  means light temp move light*temp
        light*move temp*move light*temp*move;
run;
proc sort;
  by light;
run;
proc plot;
  plot ybar*move=temp;
  by light;
run;
```

Figure 13.22. Analysis of variance for three-way factorial

```
                    General Linear Models Procedure
                       Class Level Information

          Class     Levels     Values

          LIGHT        2        high low

          TEMP         3        cold luke warm

          MOVE         2        flowing still


          Number of observations in data set = 36
```

Dependent Variable: Y

Source	DF	Sum of Squares	F Value	Pr > F
Model	11	3.00240000	11.87	0.0001
Error	24	0.55199994		
Corrected Total	35	3.55439994		

	R-Square	C.V.	Y Mean
	0.844700	8.766329	1.73000000

Source	DF	Type III SS	F Value	Pr > F
LIGHT	1	0.27040000	11.76	0.0022
TEMP	2	1.33755000	29.08	0.0001
MOVE	1	0.42250000	18.37	0.0003
LIGHT*TEMP	2	0.09965000	2.17	0.1365
LIGHT*MOVE	1	0.24010000	10.44	0.0036
TEMP*MOVE	2	0.44705000	9.72	0.0008
LIGHT*TEMP*MOVE	2	0.18515000	4.03	0.0311

The analysis of variance table is given in Figure 13.22, and the plots are shown in Figures 13.23 and 13.24. (Note that the printout of the reconstructed data set is not shown, and neither are the cell means computed by the program in Figure 13.21. Additionally, since the design is balanced and the Type I and Type III analyses are identical, Figure 13.22 shows only the Type III sums of squares and F-tests.)

Figure 13.23. Interaction between temperature and movement at low light condition

```
-------------------------- LIGHT=low --------------------------

          Plot of YBAR*MOVE.   Symbol is value of TEMP.

     YBAR |
          |
     2.0 +   w
          |
          |
          |                                                 l
          |
          |
     1.8 +
          |   l
          |
          |
          |
          |
     1.6 +
          |                                                 w
          |                                                 c
          |
          |
          |
     1.4 +
          |
          |
          |
          |
     1.2 +
          |
          |
          |
        · |   c
          |
     1.0 +
          |
         ---+-----------------------------------------+--
          flowing                                   still

                              MOVE

NOTE: 12 obs hidden.
```

Figure 13.24. Interaction between temperature and movement at high light condition

```
-------------------------- LIGHT=high --------------------------

          Plot of YBAR*MOVE.   Symbol is value of TEMP.

     YBAR |
          |
      2.2 +
          |
          |                                                    w
          |
      2.1 +
          |
          |
          |                                                    l
      2.0 +
          |
          |
          |
      1.9 +
          |
          |   w                                                c
          |
      1.8 +
          |
          |
          |
      1.7 +
          |   l
          |
          |
      1.6 +
          |
          |
          |
      1.5 +
          |
          |
          |
      1.4 +
          |
          |   c
          |
      1.3 +
          |
          ---+----------------------------------------------+--
          flowing                                        still

                               MOVE
NOTE: 12 obs hidden.
```

Note that two interaction plots are necessary--one graphing cell means against movement for low light condition and the other for high light condition. Each of these plots enables you to look at the nature of the interaction between temperature and movement. Comparing the two graphs also enables you to explore whether or not there is any *three-way inter-action:* does the interaction between temperature and movement depend on whether light is at high or low level? Apparently, it does. There appears to be interaction between temperature and movement if light is at low level, as indicated by nonparallelism of the lines for the three temperature levels. But for high light intensity, the lines relating mean response to movement appear to be parallel for the three temperature levels. The significant ($p = 0.0319$) *F*-test for the three-way interaction indicates strong evidence for this differential effect.

In the presence of a three-way interaction, the next step is to analyze the *two-way interactions at each of the levels of the third factor*. To test whether there is a significant interaction between temperature and movement at low light, we could refer back to the interaction sum of squares from Figure 13.13:

$$SS[TEMP*MOVE] = 0.6013,$$

since the analysis of the two-factor study was for low light levels. But what if no two-way analysis had been performed? After all, if this is in reality a three-factor study, then no two-way analysis would have been performed.

First, look at the Class Levels Information in Figure 13.22. Since LIGHT was entered first in the MODEL statement, SAS will list the cells in this order:

111, 112, 121, 122, 131, 132, 211, 212, 221, 222, 231, 232

or

HCF, HCS, HLF, HLS, HWF, HWL, LCF, LCS, LLF, LLS, LWF, LWS.

Now consider the interaction between temperature and movement at low light level. One way to write this two-way interaction is

$$\mu_{211} - \mu_{212} - (\mu_{231} - \mu_{232})$$

and

$$\mu_{221} - \mu_{222} - (\mu_{231} - \mu_{232}),$$

where the first subscript indicates the level of LIGHT, the second indicates the level of TEMP, and the third indicates the level of MOVE.

Now write out the expression for μ_{211}:

$$\mu_{211} = \mu + \alpha_2 + \beta_1 + \gamma_1 + (\alpha\beta)_{21} + (\alpha\gamma)_{21} + (\beta\gamma)_{11} + (\alpha\beta\gamma)_{211}.$$

Expressions for the other expected values can be written in a similar fashion. Performing the indicated subtractions, we see that the first expression can be written as

$$(\beta\gamma)_{11} - (\beta\gamma)_{12} - (\beta\gamma)_{31} + (\beta\gamma)_{32} + (\alpha\beta\gamma)_{211} - (\alpha\beta\gamma)_{212} - (\alpha\beta\gamma)_{231} + (\alpha\beta\gamma)_{232}$$

and the second as

$$(\beta\gamma)_{21} - (\beta\gamma)_{22} - (\beta\gamma)_{31} + (\beta\gamma)_{32} + (\alpha\beta\gamma)_{221} - (\alpha\beta\gamma)_{222} - (\alpha\beta\gamma)_{231} + (\alpha\beta\gamma)_{232}$$

Then the SAS code to test for interaction between temperature and movement at low light level is

```
contrast 'inter low' temp*move 1 -1 0 0 -1 1
            light*temp*move 0 0 0 0 0 0 1 -1 0 0 -1 1,
                   temp*move 0 0 1 -1 -1 1
            light*temp*move 0 0 0 0 0 0 0 0 1 -1 -1 1;.
```

Similarly, the test for interaction between temperature and movement at high light intensity is

```
contrast 'inter high' temp*move 1 -1 0 0 -1 1
         light*temp*move 1 -1 0 0 -1 1 0 0 0 0 0 0,
                     temp*move 0 0 1 -1 -1 1
         light*temp*move 0 0 1 -1 -1 1;.
```

Figure 13.25 shows the output from these two contrasts. As indicated by the interaction plots, there is a temperature by movement interaction if light is at low intensity, but not if light is at high intensity. That the two-factor interaction behaves differently at the two levels of the third factor says that there is a three-factor interaction.

Figure 13.25. Contrasts to test for MOVE*TEMP interaction at the two levels of LIGHT

```
                    General Linear Models Procedure

Dependent Variable: Y

Contrast               DF        Contrast SS      F Value      Pr > F

inter low               2        0.60130000         13.07      0.0001
inter high              2        0.03090000          0.67      0.5202
```

Since at low light intensity there is a significant interaction between temperature and movement, the effects of temperature must be analyzed separately at both of the levels of movement, and the effects of movement must be investigated separately at each of the temperature levels, as we saw in the last section. For example, to compare the effects of flowing and still cold water at low light, we want to compare

$$\mu_{211} - \mu_{212} =$$
$$\gamma_1 - \gamma_2 + (\alpha\gamma)_{21} - (\alpha\gamma)_{22} + (\beta\gamma)_{11} - (\beta\gamma)_{12} + (\alpha\beta\gamma)_{211} - (\alpha\beta\gamma)_{212}.$$

The contrast to make this comparison is

```
contrast 'move in cold low' move 1 -1
    light*move 0 0 1 -1 temp*move 1 -1 0 0 0 0
    light*temp*move 0 0 0 0 0 0 1 -1 0 0 0 0;
```

and Figure 13.26 shows the output. Note that the contrast sum of squares is the same as in Figure 13.17. The value of the F-statistic is different because its denominator in Figure 13.26 is the error mean square from the three-factor study.

Figure 13.26. Contrast of still vs. flowing cold water for low light level

General Linear Models Procedure				
Dependent Variable: Y				
Contrast	DF	Contrast SS	F Value	Pr > F
move in cold low	1	0.33135000	14.41	0.0009

Since there is no significant interaction between temperature and movement under high light intensity, analysis of temperature and movement effects under high light intensity can be accomplished as a main-effects analysis. To test whether there are significant temperature effects under high light we might compare

$$\mu_{111} + \mu_{112} - (\mu_{131} + \mu_{132})$$

and

$$\mu_{121} + \mu_{122} - (\mu_{131} + \mu_{132}).$$

The first expression may be written as

$$2\beta_1 - 2\beta_3 + 2(\alpha\beta)_{11} - 2(\alpha\beta)_{13} + (\beta\gamma)_{11} + (\beta\gamma)_{12} - (\beta\gamma)_{31} - (\beta\gamma)_{32}$$
$$+ (\alpha\beta\gamma)_{111} + (\alpha\beta\gamma)_{112} - (\alpha\beta\gamma)_{131} - (\alpha\beta\gamma)_{132}$$

and the second as

$$2\beta_2 - 2\beta_3 + 2(\alpha\beta)_{12} - 2(\alpha\beta)_{13} + (\beta\gamma)_{21} + (\beta\gamma)_{22} - (\beta\gamma)_{31} - (\beta\gamma)_{32}$$
$$+ (\alpha\beta\gamma)_{121} + (\alpha\beta\gamma)_{122} - (\alpha\beta\gamma)_{131} - (\alpha\beta\gamma)_{132}$$

The contrast

```
contrast 'temps in high'
    temp 2 0 -2
          light*temp 2 0 -2 0 0 0
          temp*move 1 1 0 0 -1 -1
          light*temp*move 1 1 0 0 -1 -1 0 0 0 0 0 0,
    temp 0 2 -1
          light*temp 0 2 -2 0 0 0
          temp*move 0 0 1 1 -1 -1
          light*temp*move  0 0 1 1 -1 -1;
```

produces the output in Figure 13.27. Other appropriate contrasts can be written to compare mean effects of temperature levels under high light intensity.

Figure 13.27. Comparing temperature levels under high light intensity

General Linear Models Procedure				
Dependent Variable: Y				
Contrast	DF	Contrast SS	F Value	Pr > F
temps in high	2	0.46630000	10.14	0.0006

Similarly, to compare the effects of flowing and still water at high light intensity, the contrast is

$$\mu_{111} + \mu_{121} + \mu_{131} - (\mu_{112} + \mu_{122} + \mu_{132})$$

or

$$3\gamma_1 - 3\gamma_2 + 3(\alpha\gamma)_{11} - 3(\alpha\gamma)_{12}$$
$$+ (\beta\gamma)_{11} + (\beta\gamma)_{21} + (\beta\gamma)_{31} - (\beta\gamma)_{12} - (\beta\gamma)_{22} - (\beta\gamma)_{32}$$
$$+ (\alpha\beta\gamma)_{111} + (\alpha\beta\gamma)_{121} + (\alpha\beta\gamma)_{131} - (\alpha\beta\gamma)_{122} - (\alpha\beta\gamma)_{122} - (\alpha\beta\gamma)_{132}$$

and the contrast statement

```
contrast 'move in high'
         move 3 -3
         light*move 3 -3 0 0
         temp*move 1 -1 1 -1 1 -1
         light*temp*move 1 -1 1 -1 1 -1 0 0 0 0 0 0;
```

produce the output in Figure 13.28. There is a significant difference in the effects of still and flowing water under the high light level. An ESTIMATE statement in place of a CONTRAST statement would give an estimate of the different effects of flowing versus still water.

Figure 13.28. Contrasting flowing vs. still water effects under high light intensity

```
                  General Linear Models Procedure

Dependent Variable: Y

Contrast               DF      Contrast SS     F Value     Pr > F

move in high            1      0.64980000       28.25      0.0001
```

13.6 Factorial Analysis when Subclass Numbers are Unequal

Example 13.16. To study the effects of two methods of fertilization (a_1, a_2) and three methods of irrigation (b_1, b_2, b_3) on the yield (bushels per acre) on a variety of corn., four plots were randomized to receive each of the six treatment combinations. However, the plants in two plots receiving a_1b_3 and one plot receiving a_2b_3 were destroyed before harvest time.

The ANOVA procedure can be used only for balanced or unbalanced single-factor studies or for balanced multifactor data. Thus, the GLM procedure is used to analyze these data. Figure 13.29 shows the SAS program. The code, with the exception of the LSMEANS command, looks the same as for balanced multifactor studies. (We will discuss the purpose of the LSMEANS shortly.) Examination of the output from the program in

Figure 13.29. Program to analyze unbalanced two-factor data

```
options ls=70 ps=54;
/*2 x 3 unbalanced factorial*/
data;
  f: input fert $ irrig$ @;
  y: input yield @@;
  if yield = -1 then go to f;
  if yield = -2 then stop;
  else output;
  go to y;
lines;
F1 I1 17.0 18.1 17.5 16.9 -1
F1 I2 16.8 15.3 14.6 16.1 -1
F1 I3 22.1 23.4 -1
F2 I1 18.3 17.8 16.8 18.9 -1
F2 I2 14.9 15.8 14.2 14.8 -1
F2 I3 24.2 26.2 25.1 -2
;
run;
proc glm;
  class fert irrig;
  model yield = fert irrig fert*irrig;
  output p=ybar;
  means fert irrig fert*irrig;
  means fert irrig/lsd cldiff;
  estimate 'f2 vs f1' fert -3 3
      fert*irrig -1 -1 -1 1 1 1/divisor=3;
  estimate 'i3 vs i1' irrig -2 0 2
      fert*irrig -1 0 1 -1 0 1/divisor=2;
  estimate 'i3 vs i2' irrig 0 -2 2
      fert*irrig 0 -1 1 0 -1 1/divisor=2;
  estimate 'i2 vs i1' irrig -2 2 0
      fert*irrig -1 1 0 -1 1 0/divisor=2;
  lsmeans fert irrig fert*irrig/pdiff;
run;
proc plot;
  plot ybar*irrig=fert;
run;
```

Figure 13.30 shows a significant Model F, but when the Model sum of squares is broken down into effects of fertilizer, irrigation, and their interaction, we see that the Type I and Type III sums of squares and F-

Figure 13.30. GLM analysis of unbalanced two-factor factorial

```
                    General Linear Models Procedure
                      Class Level Information

              Class     Levels     Values

              FERT        2        F1 F2

              IRRIG       3        I1 I2 I3


         Number of observations in data set = 21
```

General Linear Models Procedure

Dependent Variable: YIELD

Source	DF	Sum of Squares	F Value	Pr > F
Model	5	257.56142857	75.93	0.0001
Error	15	10.17666667		
Corrected Total	20	267.73809524		

	R-Square	C.V.	YIELD Mean
	0.961990	4.495121	18.3238095

Source	DF	Type I SS	F Value	Pr > F
FERT	1	5.64573160	8.32	0.0113
IRRIG	2	244.24678671	180.01	0.0001
FERT*IRRIG	2	7.66891026	5.65	0.0148

Source	DF	Type III SS	F Value	Pr > F
FERT	1	2.68015152	3.95	0.0654
IRRIG	2	228.17006410	168.16	0.0001
FERT*IRRIG	2	7.66891026	5.65	0.0148

values are no longer identical, because when the design is unbalanced, these effects are no longer orthogonal.

As was the case in multiple regression, the Type I sums of squares show a sequential partitioning of the model sum of squares, according to the order in which the variables are listed in the model statement. In order for this analysis to have meaning, then, there must be some rational ordering of the factorial effects. Each effect is adjusted for the effects of the factors that preceded it in the model statement. Thus, if the model statement is written, as in Figure 13.29, as

```
model yield = fert irrig fert*irrig;
```

then SS[IRRIG] measures the amount of variation explained by differences in the fertilizers, SS[FERT] is really SS[FERT | IRRIG] and measures the *additional* amount of variation explained by differences in irrigation treatments, after effects of different fertilizers have been taken into account, and SS[FERT*IRRIG] is really SS[FERT*IRRIG | FERT,IRRIG] and measures how much additional variation is explained by the interaction, after the two main effects have explained all the variation they can. These three sums of squares will sum to the Model *SS,* but it is unlikely that they give the information necessary for analysis of these data, and of course, changing the order of the variables in the model statement will change the analysis.

In addition to Type I sums of squares and *F*-values, the GLM procedure provides by default the Type III analysis. As was the case in multiple regression, the Type III, or *partial*, sums of squares measure the additional amount of variation explained by each term, adjusted for all the other terms in the model. It is generally felt that these are the appropriate sums of squares on which to base *F*-tests for unbalanced multifactor studies, just as they were in multiple regression analyses. However, if the unbalance is such that all the data in a cell are missing, then the Type IV analysis, also provided by GLM, should be requested.

The Type III analysis in Figure 13.30 indicates a significant interaction, significant irrigation effects, and probably also significant fertilizer effects. The graph in Figure 13.31 shows that, on the average, yields are higher for Fertilizer 2 than for Fertilizer 1, that Irrigation Method 3 seems to produce the highest yields, while Method 2 produces the lowest, and that the difference between the two fertilizers becomes more pronounced as one moves from Irrigation Method 1 to 2 to 3.

Figure 13.31. Plot of cell means for fertilizer and irrigation effects

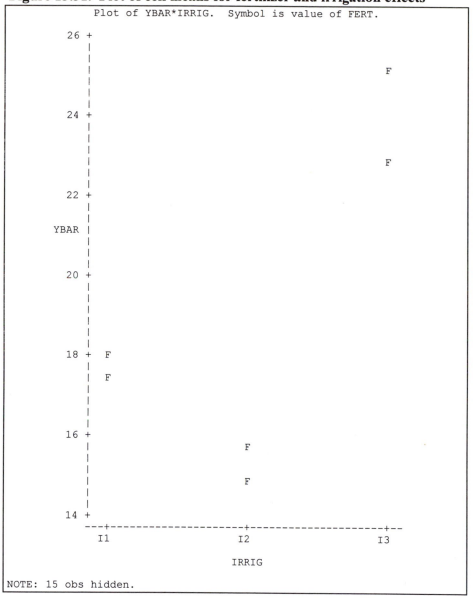

```
            Plot of YBAR*IRRIG.   Symbol is value of FERT.
     26 +
        |
        |
        |                                               F
        |
        |
        |
     24 +
        |
        |
        |
        |                                               F
        |
        |
     22 +
        |
        |
YBAR    |
        |
        |
        |
     20 +
        |
        |
        |
        |
        |
        |
     18 +   F
        |
        |   F
        |
        |
        |
     16 +
        |                          F
        |
        |
        |                          F
        |
        |
     14 +
        ---+--------------------+--------------------+--
           I1                   I2                   I3

                               IRRIG

NOTE: 15 obs hidden.
```

Recall that in the presence of a significant interaction, it is not appropriate to compare the two fertilizers, averaging over irrigation levels or to compare the three irrigation levels averaging over fertilizers, since the way fertilizer affects yield depends on the irrigation treatment, and vice versa. However, both of the main effects are strong enough to be seen even in the presence of interaction in this case. So, for purposes of illustration, let's look at the tests of pairwise differences in factor level means that result from the MEANS/LSD or the ESTIMATE commands in the GLM procedure. Output from the MEANS/LSD is shown (in confidence interval form) in Figure 13.32. We expect them to be the same differences as arrived at by writing a set of contrasts to make all pairwise comparisons, as shown in Figure 13.33, but surprisingly, that is not the case. The problem is that when the design is unbalanced, the difference in factor level means does not measure the difference in the effects of the factors, but is contaminated by some of the effects of the other factor. The difference in the means must be *adjusted* to remove the unwanted effects. You will notice in the SAS LOG file, that whenever pairwise comparisons are requested, SAS prints a message: "NOTE: Means from the MEANS statement are not adjusted for other terms in the model. For adjusted means, use the LSMEANS statement. "LSMEANS" stands for "least squares means", but this is terminology unique to SAS. Using the LSMEANS command, as in the program in Figure 13.29, will produce the adjusted means for each factor level and for each treatment combination, along with a test of the null hypothesis that each of the corresponding population means is zero. Additionally, it is possible to request all pairwise comparisons among adjusted means. Including TDIFF after the slash in the LSMEANS command will show the value of the t-statistic for testing the difference, and PDIFF will give the p-value associated with that t-value. Figure 13.34 shows the output from LSMEANS with the p-values requested.

Figure 13.32. Pairwise comparison of (unadjusted) factor level means using MEANS command

```
                 T tests (LSD) for variable: YIELD

      NOTE: This test controls the type I comparisonwise error rate
            not the experimentwise error rate.

          Alpha= 0.05  Confidence= 0.95  df= 15  MSE= 0.678444
                     Critical Value of T= 2.13145

   Comparisons significant at the 0.05 level are indicated by '***'.

                         Lower     Difference     Upper
              FERT     Confidence    Between    Confidence
           Comparison    Limit       Means        Limit

        F2   - F1        0.2711      1.0382       1.8053    ***

        F1   - F2       -1.8053     -1.0382      -0.2711    ***


                 T tests (LSD) for variable: YIELD

      NOTE: This test controls the type I comparisonwise error rate
            not the experimentwise error rate.

          Alpha= 0.05  Confidence= 0.95  df= 15  MSE= 0.678444
                     Critical Value of T= 2.13145

   Comparisons significant at the 0.05 level are indicated by '***'.

                         Lower     Difference     Upper
              IRRIG    Confidence    Between    Confidence
           Comparison    Limit       Means        Limit

        I3   - I1        5.5366      6.5375       7.5384    ***
        I3   - I2        7.8866      8.8875       9.8884    ***

        I1   - I3       -7.5384     -6.5375      -5.5366    ***
        I1   - I2        1.4722      2.3500       3.2278    ***

        I2   - I3       -9.8884     -8.8875      -7.8866    ***
        I2   - I1       -3.2278     -2.3500      -1.4722    ***
```

Figure 13.33. **Results from writing contrasts to make all pairwise comparisons**

```
Dependent Variable: YIELD

                              T for H0:     Pr > |T|   Std Error of
Parameter          Estimate   Parameter=0              Estimate

f2 vs f1          0.73888889         1.99    0.0654    0.37175482
i3 vs i1          6.29583333        13.24    0.0001    0.47555036
i3 vs i2          8.64583333        18.18    0.0001    0.47555036
i2 vs i1         -2.35000000        -5.71    0.0001    0.41183870
```

Comparing the output in Figures 13.33 and 13.34 shows that the estimates of the differences in Figure 13.33 are, indeed, made using adjusted means. If the objective is simply to compare factor level means, then it is probably more convenient to use the LSMEANS command than to write CONTRAST or ESTIMATE statements. However, in the presence of significant interaction, or to make comparisons other than pairwise comparisons, CONTRAST and ESTIMATE should be used to compare adjusted means. It is certainly difficult to make much sense of the pairwise comparisons of adjusted cell means in Figure 13.34.

Finally, note that in the LSMEANS output, SAS gives you a warning about making after-the-fact tests: "NOTE: To ensure overall protection level, only probabilities associated with pre-planned comparisons should be used."

Figure 13.34. LSMEANS to compare adjusted means

```
                    General Linear Models Procedure
                         Least Squares Means

              FERT           YIELD      Pr > |T| H0:
                             LSMEAN    LSMEAN1=LSMEAN2

              F1          18.6083333        0.0654
              F2          19.3472222


        IRRIG          YIELD   Pr > |T| H0: LSMEAN(i)=LSMEAN(j)
                       LSMEAN   i/j    1       2       3

        I1         17.6625000   1    .     0.0001  0.0001
        I2         15.3125000   2  0.0001    .     0.0001
        I3         23.9583333   3  0.0001  0.0001    .

NOTE: To ensure overall protection level, only probabilities
      associated with pre-planned comparisons should be used.


            FERT    IRRIG           YIELD    LSMEAN
                                    LSMEAN   Number

            F1      I1          17.3750000     1
            F1      I2          15.7000000     2
            F1      I3          22.7500000     3
            F2      I1          17.9500000     4
            F2      I2          14.9250000     5
            F2      I3          25.1666667     6

            Pr > |T| H0: LSMEAN(i)=LSMEAN(j)

       i/j    1        2       3       4       5       6
       1    .      0.0115  0.0001  0.3392  0.0008  0.0001
       2  0.0115     .     0.0001  0.0015  0.2032  0.0001
       3  0.0001  0.0001     .     0.0001  0.0001  0.0058
       4  0.3392  0.0015  0.0001     .     0.0001  0.0001
       5  0.0008  0.2032  0.0001  0.0001     .     0.0001
       6  0.0001  0.0001  0.0058  0.0001  0.0001     .

NOTE: To ensure overall protection level, only probabilities
      associated with pre-planned comparisons should be used.
```

13.7 Nested Factors

Example 13.19. The amount of readily soluble phosphorus in a large number of soil samples was to be determined in a laboratory that employed six technicians. Three technicians worked in the morning shift and the other three worked in the evening shift. In a study to determine whether the measured values of phosphorus (lb/acre) were affected by (1) the time of day (A.M. and P.M.), and (2) the technician making the measurement, 24 identical specimen samples were assigned to the six technicians at random, in such a way that each received four samples. Each technician was asked to analyze independently the four assigned samples.

In order to indicate that levels of factor *B* are nested within levels of factor *A,* use the following notation in the model statement in either the ANOVA (if the design is balanced) or the GLM procedure:

```
model y = A B(A);.
```

If a third factor, factor *C,* is nested within levels of factor *B,* this is indicated by

```
model y = A B(A) C(B A);.
```

It is possible to have a combination of nested and crossed factors. Nested effects are specified by following a main effect or crossed effects with a list of one or more class variables enclosed in parentheses. These effects are nested within the effects enclosed in parentheses. For example, the model statement

```
model A B(A) C*D(A B);
```

would indicate that factor *B* is nested within levels of factor *A,* and then that *C*D* combinations are nested within levels of factor *B.* The distinguishing feature of a nested factor in a model statement is that it never appears as a main effect. Thus, in the two-factor model above, since *B* never appears alone, it is assumed to be nested in A. In the three-factor example, neither *B* nor *C* appears alone. In the last example, neither of *B, C,* nor *D* appears alone.

The SAS program to analyze the data of Example 13.19 is shown in Figure 13.35. Since the design is balanced, either the ANOVA or the GLM procedure can be used: Figure 13.35 shows the ANOVA procedure.

Figure 13.35. ANOVA procedure for analyzing a two-factor nested design

```
options ls=70 ps=54;
/*Two nested factors*/
data;
  input shift $@@;
    do j = 1 to 3;
      input tech @@;
        do k = 1 to 4;
          input phos @@;
            output;
        end;
    end;
  drop j; drop k;
lines;
AM 1 42 44 43 44 2 43 44 45 42 3 47 46 47 43
PM 1 50 49 52 50 2 49 48 49 47 3 47 51 46 48
;
run;
proc anova;
  class shift tech;
  model phos = shift tech(shift);
  means shift tech(shift);
run;
```

Note that since the three technicians on the morning shift are not the same three persons as the three technicians on the evening shift, it is often advisable to number them 1 through 6 instead of 1 through 3 in the morning and 1 through 3 in the evening. This will prevent any confusion as to whether the factors are nested or crossed. SAS would have had no trouble reading the input statement or executing the analysis had the data been entered as

```
AM 1 42 44 43 44 2 43 44 45 42 3 47 46 47 43
PM 4 50 49 52 50 5 49 48 49 47 6 47 51 46 48
```

where the boldface entries indicate changes in the data from Figure 13.35.

The analysis of the nested model is shown in Figure 13.36 and the means and standard deviations of the phosphorous levels for each of the factor levels are shown in Figure 13.37. We see that the readings made by the PM shift are, on the average, significantly higher than those made during the AM shift. Additionally, within any shift, the technicians also differ significantly from each other.

Figure 13.36. Analysis of nested two-factor model

```
                      Analysis of Variance Procedure

Dependent Variable: PHOS

Source                   DF    Sum of Squares    F Value     Pr > F

Model                     5     158.00000000      14.22      0.0001

Error                    18      40.00000000

Corrected Total          23     198.00000000

                R-Square              C.V.             PHOS Mean

                0.797980             3.205832          46.5000000


Source                   DF        Anova SS      F Value     Pr > F

SHIFT                     1     130.66666667      58.80      0.0001
TECH(SHIFT)               4      27.33333333       3.08      0.0429
```

Suppose we wanted a confidence interval estimate on the difference between the mean readings on the AM and PM shifts. *Provided the data are analyzed with GLM instead of with ANOVA*, the command

```
estimate 'shift' shift -3 3
        tech(shift) -1 -1 -1 1 1 1/divisor = 3;.
```

produces the estimate in Figure 13.38. To figure out what the constants in the ESTIMATE statement are, consider first that we want to estimate

$$(\mu_{11} + \mu_{12} + \mu_{13}) / 3 - (\mu_{21} + \mu_{22} + \mu_{23}) / 3,$$

Figure 13.37. Means and standard deviations at each of the factor levels

```
                    Analysis of Variance Procedure

           Level of              -------------PHOS------------
           SHIFT        N         Mean                 SD

           AM          12      44.1666667          1.74945879
           PM          12      48.8333333          1.74945879

   Level of   Level of          -------------PHOS------------
   TECH       SHIFT      N         Mean                 SD

   1          AM         4      43.2500000          0.95742711
   2          AM         4      43.5000000          1.29099445
   3          AM         4      45.7500000          1.89296945
   1          PM         4      50.2500000          1.25830574
   2          PM         4      48.2500000          0.95742711
   3          PM         4      48.0000000          2.16024690
```

Figure 13.38. Estimate of the difference in mean readings on AM and PM shifts

Parameter	Estimate	T for H0: Parameter=0	Pr > |T|	Std Error of Estimate
shift	4.66666667	7.67	0.0001	0.60858062

or, equivalently, one-third of

$$(\mu_{11} + \mu_{12} + \mu_{13}) - (\mu_{21} + \mu_{22} + \mu_{23}) .$$

Writing, for example,

$$\mu_{21} = \mu + \alpha_2 + \beta_{1(2)} ,$$

and performing the indicated sums and differences, we obtain

$$3\alpha_1 - 3\alpha_2 + \beta_{1(1)} + \beta_{2(1)} + \beta_{3(1)} - \beta_{1(2)} - \beta_{2(2)} - \beta_{3(2)} .$$

(Actually, the difference in Figure 13.38 is for PM minus AM instead of AM minus PM as indicated above, in order to obtain a positive difference.) Note that the order of the subscripts on the nested factor are

$$1(1), 2(1), 3(1), 1(2), 2(2), \text{ and } 3(2),$$

so that the level of *A* is held constant while the level of *B* varies. In a similar fashion, if an (after the fact) comparison is to be made, *for the evening shift*, between Technician 1 on the one hand and the average of Technicians 2 and 3 on the other, the contrast statement *in GLM* would be

```
contrast '4 vs 5&6' tech(shift) 0 0 0 1 -.5 -.5;,
```

which produces the output in Figure 13.39.

Figure 13.39. Contrasting the first with the other two PM technicians

Contrast	DF	Contrast SS	F Value	Pr > F
4 vs 5&6	1	12.04166667	5.42	0.0318

14

Analysis of Variance Models with Random Effects

14.1 Introduction

In the ANOVA and GLM procedures, SAS assumes fixed effects factors unless you tell it otherwise. This chapter shows how to specify that one or more factors are random effects factors. The table of expected mean squares produced by the RANDOM statement may be used to indicate how to estimate the variance components of interest. The TEST command in GLM allows the user to specify the numerator and denominator of an *F*-statistic for testing both random and fixed effects factors.

The NESTED procedure may be used in a nested design in which all factors are random effects factors. The MIXED procedure is appropriate in multifactor studies in which some factors are fixed effects factors and others are random.

14.2 The One-Way Classification Model with Random Effects

Example 14.1. To assess the batch to batch variation in the protein content of 250-ml cartons of pasteurized buttermilk produced by a dairy farm, four batches of cartons were randomly selected from the production line. From each batch, protein contents were determined for three randomly selected cartons.

In this study, the experimental factor is the batch because the objective was to see how the protein contents of the buttermilk cartons vary from one batch to another. It was impossible to make protein content measurements for samples selected from every batch of production, and so the investigators decided to confine the study to four randomly selected batches from the population of all batches. In other words, the investigators decided to restrict their study to $t = 4$ treatments (batches) selected from a large (infinite) number of treatments.

Even though only four treatments were included for observation, the objective of the study extended beyond the effects of these treatments. The investigators wanted to study the carton-to-carton variability of protein content over all batches of cartons produced by the dairy farm.

The analysis of variance table for a one-way random effects model is identical to that for a one-way fixed effects model, even though the hypotheses being tested differ. In the random effects model we are testing

$$H_0: \sigma_T^2 = 0$$

This hypothesis is tested with the F-statistic,

$$F = \frac{MS[T]}{MS[E]}$$

just as the hypothesis

$$H_0: \tau_1 = \tau_2 = \cdots = \tau_t$$

was tested in the fixed effects model. Either the ANOVA or the GLM procedure will produce the appropriate analysis of variance table:

```
proc anova;    (or proc glm;)
    class batch;
    model protein = batch;
```

However, if the null hypothesis of no variation due to batches is rejected, then the question becomes, "How large is the batch-to-batch variation?" and what is desired is an estimate of σ_T^2. The table of expected mean squares for the random effects model, shown in Rao's Box 14.2, indicates how this estimate is made.

If you analyze the data using GLM, you can use a RANDOM statement to request that a table of expected mean squares be printed. (This is not available in the ANOVA procedure.) SAS will compute the numerical values for the coefficient n_0 in the expected mean squares, and then you can do the necessary algebra to estimate the batch-to-batch variance. Figure 14.1 shows the SAS code.

Figure 14.1. SAS code to analyze one-way random effects model

```
options ls=70 ps=54;
/*Example 14.1. One-Way Random Effects*/
data;
   input batch @;
   do i = 1 to 3;
      input protein @;
      output;
   end;
   drop i;
cards;
1 3.42 3.41 3.57
2 3.05 3.14 3.23
3 3.23 3.48 3.37
4 3.46 3.59 3.23
;
run;
proc glm;
   class batch;
   model protein = batch;
   random batch;
run;
```

Figure 14.2 shows the output from GLM. The class levels information and the analysis of variance table look as they do for the fixed effects model. The *p*-value for the *F*-test is $p = 0.0545$, probably indicating significant batch-to-batch variation in protein content.

Figure 14.2. SAS output for one-way random effects model

```
                    General Linear Models Procedure
Dependent Variable: PROTEIN
```

Source	DF	Sum of Squares	F Value	Pr > F
Model	3	0.19103333	3.91	0.0545
Error	8	0.13013333		
Corrected Total	11	0.32116667		

R-Square	C.V.	PROTEIN Mean
0.594811	3.809084	3.34833333

Source	DF	Type I SS	F Value	Pr > F
BATCH	3	0.19103333	3.91	0.0

Source	DF	Type III SS	F Value	Pr > F
BATCH	3	0.19103333	3.91	0.0545

If this batch-to-batch variation is not zero, then to estimate how large it is, refer in Figure 14.3 to the Type III Expected Mean Square. This table results from the

```
                        random batch;
```
statement.

Figure 14.3. Table of expected mean squares for one-way random effects model

```
                    General Linear Models Procedure
```

Source	Type III Expected Mean Square
BATCH	Var(Error) + 3 Var(BATCH)

It is understood that there is another line in the table in Figure 14.3, namely,

Source	Type III Expected Mean Square
ERROR	Var(Error)

Thus, to estimate Var(BATCH), use Equation. 14.5 in Rao:

$$\hat{\text{Var}}(\text{BATCH}) = [\text{MS}(\text{BATCH}) - \text{MS}(\text{ERROR})] / 3$$
$$= (0.0636778 - 0.0162667)/3$$
$$= 0.0158037.$$

14.3 Factorial Models with Random Effects Factors

Examples 14.12 and 14.15. A 3 × 4 factorial study was designed to investigate the effects of the fertilization method (factor *A*, with three levels) and seeding rate (factor *B*, with levels varying in the range 10-20 kg per acre) on the yield of sugar beets. Because the investigators were interested in evaluating the variation in yield that can result from varying seeding rate, they decided to conduct a study by selecting a random sample of four seeding rates from the range of interest. The study was conducted in 36 quarter-acre plots. Each treatment combination was replicated in three plots. (Yield is tons per acre.)

Figure 14.4 shows the SAS code to analyze this two-factor, mixed model. By default, SAS assumes fixed effects models and will print an analysis of variance table in which *F*-statistics for all main effects and interactions are formed with MS[E] as the denominator, as shown in Figure 14.5.

Using the GLM procedure with a RANDOM statement, you can request that the table of expected mean squares be printed (see Figure 14.6), showing you how to form the appropriate *F*-statistics in random or mixed models and how to estimate the variances of the random effects factors. Note that in the program shown in Figure 14.4, both factor *B* and the interaction between the fixed and random effects factors are declared to be random effects. As was the case in the table of expected mean squares for a

Figure 14.4. **SAS code to analyze two-way mixed model (crossed factors)**

```
options ls=70 ps=54;
/*Examples 14.12 and 14.15.  Two-Way Crossed
Factors; Mixed Model*/
data;
   input A B @;
     do i = 1 to 3;
        input yield @;
        output;
     end;
   drop i;
   cards;
1 1 15.71 16.02 15.90
1 2 16.21 16.36 16.33
1 3 17.32 17.03 17.22
1 4 17.54 17.82 17.62
2 1 17.83 17.45 16.96
2 2 17.68 17.70 17.52
2 3 17.95 18.01 18.41
2 4 18.08 18.56 18.90
3 1 14.78 15.03 14.63
3 2 15.80 15.62 15.77
3 3 16.21 16.44 16.32
3 4 16.99 16.39 17.02
;
proc glm;
   class A B;
   model yield = A B A*B;
   random B A*B/test;
   test h = A B e=A*B;
   means A/lsd e=A*B;
run;
```

single random effects factor, it is understood that there is an additional line in Figure 14.6:

```
Source                Type III Expected Mean Square

ERROR                 Var(Error)
```

In the expected mean square for the fixed effects factor (factor A) is a term

Figure 14.5. Default output for two-way analysis

```
               General Linear Models Procedure
                   Class Level Information

              Class      Levels     Values

               A            3       1 2 3

               B            4       1 2 3 4


           Number of observations in data set = 36



               General Linear Models Procedure

Dependent Variable: YIELD

Source              DF     Sum of Squares     F Value      Pr > F

Model               11       39.06083056       61.30       0.0001

Error               24        1.39026667

Corrected Total     35       40.45109722

          R-Square              C.V.              YIELD Mean

          0.965631            1.427132            16.8647222


Source              DF       Type I SS        F Value      Pr > F
A                    2       24.31027222       209.83       0.0001
B                    3       13.81794167        79.51       0.0001
A*B                  6        0.93261667         2.68       0.0389

Source              DF       Type III SS       F Value      Pr > F

A                    2       24.31027222       209.83       0.0001
B                    3       13.81794167        79.51       0.0001
A*B                  6        0.93261667         2.68       0.0389
```

$$Q(A).$$

This simply indicates a quadratic function of the factor A effects; it is denoted by

$$nb\,\psi_A^2$$

Figure 14.6. Expected mean squares for the mixed model (crossed factors)

```
                    General Linear Models Procedure

Source          Type III Expected Mean Square

A               Var(Error) + 3 Var(A*B) + Q(A)

B               Var(Error) + 3 Var(A*B) + 9 Var(B)

A*B             Var(Error) + 3 Var(A*B)
```

in Rao, Table 14.1.

We see from the table of expected mean squares that the interaction effect is tested by an *F*-statistic whose numerator is MS[A*B] and whose denominator is MS[E], so that the *F*-value and *p*-value for *A*B* in Figure 14.5 are appropriate. However, according to the table of expected mean squares, both *A* and *B* main effects should be tested using an *F*-statistic whose denominator is MS[A*B], not MS[E]; thus, the results pertaining to main effects in Figure 14.5 are not appropriate.

At this point, you can either compute the *F*-values by hand and look up their *p*-values in tables of the *F*-distribution, or you can use the TEST statement in GLM to have SAS compute the appropriate *F*-statistics and *p*-values for you. Note that the TEST command in the GLM procedure is different from the TEST command in the REG procedure. TEST in GLM allows you to specify the numerator (by writing "H = ...") and the denominator (by writing "E = ...") of an *F*-statistic. In Figure 14.4,

```
test h = A B e = A*B;
```

requests that both main effects be tested using *F*-statistics with the interaction mean square as the denominators. Results of the TEST command are shown in Figure 14.7.

Figure 14.7. Using GLM's TEST command

```
Dependent Variable: YIELD

Tests of Hypotheses using the Type III MS for A*B as an error term

Source              DF        Type III SS    F Value      Pr > F

A                    2       24.31027222      78.20       0.0001
B                    3       13.81794167      29.63       0.0005
```

There is another way to have the SAS change from the default (fixed effects) analysis and test the main effects against the interaction term mean square: note the

```
/test
```

option in the RANDOM statement. If this option is included, Figure 14.8 results. You might ask why the tests in Figure 14.8 are not automatically performed when the RANDOM statement is used, and why it is necessary to have the TEST H = ... E = ... command in GLM if this /TEST option is available. The RANDOM statement in GLM obtains expected mean square values under the assumption that the interaction effects, $(\alpha\beta)_{ij}$, are independent $N(0, \sigma^2_{\alpha\beta})$.

In this example, even though the interaction between factors A and B is significant, it is nonetheless possible to tell that some level(s) of factor A give higher mean yields than others. Thus, you might want to perform all pairwise comparisons or test some other contrasts among factor A levels. The program in Figure 14.4 shows how to specify that MS[A*B] is to be used as the error term when performing Fisher's LSD pairwise comparisons. Note in Figure 14.9 that MS[E] is stated as 0.1555436, which is the interaction mean square, instead of MS[E] = 0.0579278. In order to obtain confidence interval estimates of the pairwise differences, simply writing

```
means A/lsd e=A*B cldiff;
```

results in the output in Figure 14.10.

Figure 14.8. Output from /TEST option in the RANDOM statement

```
                General Linear Models Procedure
          Tests of Hypotheses for Mixed Model Analysis of Variance

Dependent Variable: YIELD


Source: A
Error: MS(A*B)
                         Denominator    Denominator
     DF    Type III MS        DF             MS      F Value  Pr > F
      2   12.155136111         6      0.1554361111   78.2002  0.0001

Source: B
Error: MS(A*B)
                         Denominator    Denominator
     DF    Type III MS        DF             MS      F Value  Pr > F
      3   4.6059805556         6      0.1554361111   29.6326  0.0005

Source: A*B
Error: MS(Error)
                         Denominator    Denominator
     DF    Type III MS        DF             MS      F Value  Pr > F
      6   0.1554361111        24      0.0579277778    2.6833  0.0389
```

Figure 14.9. Pairwise comparisons using MS[A*B)] as an error term

```
                 T tests (LSD) for variable: YIELD

   NOTE: This test controls the type I comparisonwise error rate
            not the experimentwise error rate.

               Alpha= 0.05  df= 6  MSE= 0.155436
                    Critical Value of T= 2.45
               Least Significant Difference= 0.3938

     Means with the same letter are not significantly different.

           T Grouping          Mean       N  A

               A              17.9208     12  2

               B              16.7567     12  1

               C              15.9167     12  3
```

If you wanted to contrast the average yields at levels 1 and 2 of factor *A* to the yield at level 3, the appropriate command would have been

Figure 14.10. Estimates of pairwise differences using MS[A*B] as the error term

```
                 T tests (LSD) for variable: YIELD

  NOTE: This test controls the type I comparisonwise error rate
          not the experimentwise error rate.

      Alpha= 0.05  Confidence= 0.95  df= 6  MSE= 0.155436
                 Critical Value of T= 2.44691
             Least Significant Difference= 0.3938

Comparisons significant at the 0.05 level are indicated by '***'.

                          Lower     Difference    Upper
                A       Confidence   Between    Confidence
            Comparison     Limit      Means        Limit

          2   - 1         0.7703     1.1642       1.5580    ***
          2   - 3         1.6103     2.0042       2.3980    ***

          1   - 2        -1.5580    -1.1642      -0.7703    ***
          1   - 3         0.4462     0.8400       1.2338    ***

          3   - 2        -2.3980    -2.0042      -1.6103    ***
          3   - 1        -1.2338    -0.8400      -0.4462    ***
```

```
contrast '12vs3' A .5 .5 -1/e=A*B;
```

Annoyingly, however, GLM does not allow you to obtain an estimate of the size of this difference by simply replacing CONTRAST with ESTIMATE when the error term is specified as MS[A*B]. Doing so only results in a message that specifying the error term is not a valid option in the ESTIMATE command. Thus, any confidence intervals on contrasts other than pairwise comparisons in mixed models must be computed by hand, or the MIXED model procedure can be used.

Figure 14.11 shows how to use the MIXED model procedure to analyze these data. The commands are very much like those used in GLM, the primary difference being that *only the fixed effects factors* are listed in the MODEL statement. The random effects factors are listed in the RANDOM statement.

Figure 14.11. Two crossed factors analyzed using the MIXED model procedure

```
options ls=70 ps=54;
/*Two-Way Crossed Factors - Mixed Model*/
data a;
  input A B @;
    do i = 1 to 3;
      input yield @;
      output;
    end;
lines;

     [Data as in Figure 14.4]
;
run;
proc mixed data=a;
  class A B;
  model yield=A;
  random B A*B;
  contrast '12vs3' A .5 .5 -1;
  estimate '12vs3' A .5 .5 -1/cl;
run;
```

The CONTRAST and ESTIMATE statements in the MIXED model procedure will use the appropriate error term; in addition, the CL option in the ESTIMATE statement will result in two-sided 95% confidence limits for the contrast being computed. (The confidence coefficient can be changed by specifying ALPHA = ... after the slash.) Output from the MIXED procedure is shown in Figures 14.12 and 14.13.

Under "Class Levels Information," we see that there are three levels of factor A and four levels of factor B. The portions of the printout labeled "REML Estimation Iteration History," "Covariance Parameter Estimates (REML)," and "Model Fitting Information" refer to the procedure used to fit the model and need not concern us now. The "Test of Fixed Effects" gives the same *F*-test for factor *A* as in Figures 14.7 and 14.8.

Figure 14.12. **Output from the MIXED model procedure for two crossed factors--first page**

```
                        The MIXED Procedure

                    Class Level Information

                  Class     Levels   Values

                  A            3    1 2 3
                  B            4    1 2 3 4


              REML Estimation Iteration History

         Iteration   Evaluations     Objective     Criterion

                0            1     16.85457905
                1            1    -34.49773439    0.00000000

                  Convergence criteria met.


              Covariance Parameter Estimates (REML)

Cov Parm      Ratio        Estimate      Std Error      Z   Pr > |Z|

B          8.53657705     0.49450494    0.41798199    1.18   0.2368
A*B        0.56109140     0.03250278    0.03042860    1.07   0.2854
Residual   1.00000000     0.05792778    0.01672231    3.46   0.0005


              Model Fitting Information for YIELD

           Description                        Value

           Observations                      36.0000
           Variance Estimate                  0.0579
           Standard Deviation Estimate        0.2407
           REML Log Likelihood              -13.0761
           Akaike's Information Criterion   -16.0761
           Schwarz's Bayesian Criterion     -18.3209
           -2 REML Log Likelihood            26.1522


                  Tests of Fixed Effects

           Source    NDF   DDF   Type III F   Pr > F

             A         2     6       78.20    0.0001
```

Figure 14.13. Output from the MIXED model procedure for two crossed factors--second page

```
                   ESTIMATE Statement Results

    Parameter            Estimate     Std Error    DDF         T

    12vs3               1.42208333    0.13938979     6      10.20


                   ESTIMATE Statement Results

              Pr > |T|   Alpha      Lower       Upper

               0.0001    0.05      1.0810      1.7632


                   CONTRAST Statement Results

        Source                   NDF    DDF       F     Pr > F

        12vs3                     1      6     104.09   0.0001
```

The results of the CONTRAST and ESTIMATE statements shown in Figure 14.13 are of interest to us now. The difference between the average response in levels 1 and 2 of factor A and the response in level 3 is estimated to be 1.422, with a standard error of 0.139, giving a t-value of 10.20, which is significant at $p < 0.0001$. We are 95% confident that the difference in yields is between 1.08 and 1.76 tons/acre. Equivalently, the difference could be tested with an F-test, as shown under the "CONTRAST Statement Results" portion of Figure 14.13.

14.4 Nested Models with Random Effects Factors

Example 14.16. In a study to develop and evaluate a technique of measuring radon flux at the earth-air interface using charcoal canisters, the experimental area--a 44-acre tract of reclaimed phosphate land in Lakeland, Florida--was surveyed for gamma radiation, and a contour map showing contours of equal gamma radiation was prepared. Three contours were selected at random. Within each selected contour, two sites were selected at random. At each site, the radon flux (pCi/m^2 h) was sampled by charcoal canisters over a 30 hr period. The log (radon flux) data obtained are as follows...

Since both factors are random effects factors, we have a choice of using either the GLM procedure or the NESTED procedure to analyze these data. Note carefully, however, that the NESTED procedure can be used to analyze nested designs *only if all factors are random effects factors.* Nested

Figure 14.14. SAS program using both GLM and NESTED for a two-factor nested model with random effects factors

```
options ls=70 ps=54;
/*Example 14.16.  Two nested random factors*/
data radon;
  input contours sites @;
    do i = 1 to 3;
      input flux @;
      output;
    end;
  drop i;
cards;
1 1   0.6130 -0.4050 -0.6445
1 2   0.8303 -0.1154  1.2507
2 1   0.9450  0.3500  0.6060
2 2 -1.3205 -0.7875 -1.3548
3 1   0.6136  0.0266  0.2776
3 2 -0.4636 -0.9782 -0.1335
;
run;
proc glm;
   title 'Analysis using GLM';
   class contours sites;
   model flux = contours sites(contours);
   random contours sites(contours)/test;
run;
proc sort data=radon;
   by contours sites;
proc nested;
   title 'Analysis using NESTED';
   class contours sites;
   var flux;
run;
```

fixed effects must be analyzed using GLM, and mixed models may be analyzed using GLM or MIXED.

Figure 14.14 shows the SAS code to use both GLM and NESTED for this data set (although obviously it is redundant to use both procedures.).

Looking first at the GLM analysis, note that in order to tell SAS that levels of factor *B* are nested within, rather than crossed with, levels of factor *A*, the notation

```
sites(contours)
```

is used. Note that in a three-factor nested experiment, in which levels of factor *B* are nested within levels of factor *A*, and levels of factor *C* are nested in levels of factor *B*, the model statement would be written as

```
model y = A B(A) C(A B);
```

and would be similar for a greater number of nested factors.

By default, SAS will assume all factors are *fixed* effects unless you tell it otherwise by using the RANDOM statement. The GLM analysis for two nested random effects factors is shown in Figures 14.15-14.18. The default analysis is in Figure 14.15. Since this is the same analysis that would have been obtained if both factors had been fixed effects, we now need to look at the table of expected mean squares to determine whether the output in Figure 14.15 is appropriate. The RANDOM statement produces the table of expected mean squares in Figure 14.16. If you want SAS to compute the *F*-statistics and corresponding *p*-values in random or mixed models, then include the /TEST option in the RANDOM statement. These tests are shown in Figure 14.17.

From Figure 14.17 we see that while there is strong evidence of variation from site to site within contours, there is little evidence of variation in flux measurements from one contour to another.

Figure 14.15. GLM analysis for two nested random factors

```
                    Analysis using GLM

              General Linear Models Procedure
                 Class Level Information

            Class     Levels    Values

            CONTOURS     3      1 2 3

            SITES        2      1 2


        Number of observations in data set = 18


                 Analysis using GLM

           General Linear Models Procedure

Dependent Variable: FLUX

Source           DF    Sum of Squares    F Value    Pr > F

Model            5        7.63447712      6.57      0.0036

Error           12        2.78721480

Corrected Total 17       10.42169192

         R-Square            C.V.              FLUX Mean

         0.732556         -1256.875           -0.03834444


Source           DF      Type I SS      F Value    Pr > F

CONTOURS          2      0.84181338       1.81      0.2052
SITES(CONTOURS)   3      6.79266374       9.75      0.0015

Source           DF      Type III SS    F Value    Pr > F

CONTOURS          2      0.84181338       1.81      0.2052
SITES(CONTOURS)   3      6.79266374       9.75      0.0015
```

At this point, we would estimate the variance components due to contours and due to sites within contours. Since

Figure 14.16. Expected mean squares for two nested random factors

```
                          Analysis using GLM

                    General Linear Models Procedure

Source                    Type III Expected Mean Square

CONTOURS                  Var(Error) + 3 Var(SITES(CONTOURS))
                          + 6 Var(CONTOURS)

SITES(CONTOURS)    Var(Error) + 3 Var(SITES(CONTOURS))
```

Figure 14.17. Tests for effects of random nested factors

```
                          Analysis using GLM

                    General Linear Models Procedure
             Tests of Hypotheses for Random Model Analysis of Variance

Dependent Variable: FLUX


Source: CONTOURS
Error: MS(SITES(CONTOURS))
                          Denominator    Denominator
        DF    Type III MS        DF           MS      F Value   Pr > F
         2    0.4209066906        3    2.2642212478    0.1859   0.8392

Source: SITES(CONTOURS)
Error: MS(Error)
                          Denominator    Denominator
        DF    Type III MS        DF           MS      F Value   Pr > F
         3    2.2642212478       12     0.2322679       9.7483   0.0015
```

$$\{MS(CONTOURS) - MS[SITES(CONTOURS)]\}/6$$
$$= (0.420907 - 2.26421)/6$$
$$= -0.30722,$$

the estimate of $\sigma^2_{contours}$ is taken as zero. However, the estimate of the variance among sites within contours is given by

$$\{MS[SITES(CONTOURS)] - MS(E)\}/3$$
$$= (2.26422 - 0.23227)/3$$
$$= 0.67732.$$

Note that if CONTOURS had been a fixed effects factor, then the table of expected mean squares would have been as in Figure 14.18. One may want to test or estimate some contrasts comparing the three contours to each other. Any of these contrasts should be tested or estimated using MS[SITES(CONTOURS)] as an "error" term. Such estimates are best made using the MIXED model procedure. The SAS statements to analyze these data using the MIXED model procedure is

```
proc mixed;
    class contours sites;
    model flux = contours;
    random sites(contours);
```

with appropriate CONTRAST or ESTIMATE commands.

Figure 14.18. Expected mean squares for a mixed model

```
                        Analysis using GLM

                   General Linear Models Procedure

Source                Type III Expected Mean Square

CONTOURS              Var(Error) + 3 Var(SITES(CONTOURS)) + Q(CONTOURS)

SITES(CONTOURS)   Var(Error) + 3 Var(SITES(CONTOURS))
```

If all factors are random effects factors, then a nested structure can be analyzed using the NESTED procedure. The bottom portion of Figure 14.14 shows the code necessary to produce the analysis. Note that although a CLASS statement is required, as in GLM, instead of a MODEL statement it is only necessary to designate the dependent variable using a VAR statement. NESTED assumes the model is a fully nested one, with the hierarchical structure as indicated by the CLASS statement. That is, if factor A is listed first in the CLASS statement, followed by factor B and then factor C, NESTED assumes that factor C levels are nested within factor B levels, which are nested within factor A levels.

Either the data must be entered in the data set in the hierarchical order, or else the data set must be sorted by the CLASS variables in the order that they are given in the CLASS statement. In Figure 14.14, since the input statement lists CONTOURS first, followed by SITES, it is not really necessary to sort the data set. The

```
proc sort data=radon;
  by contours sites;
```

is included in Figure 14.14 primarily as a reminder.

Figure 14.19 shows the output from the NESTED procedure. The first information given is a table of expected mean squares, in somewhat abbreviated form. Comparing this table to the one in Figure 14.16, its meaning should be clear. Then an analysis of variance table is given, which should also be easily interpretable with reference to the GLM ANOVA table in Figure 14.15 and the tests performed in Figure 14.17.

Information on variance components follows. In the first column are mean squares for CONTOURS, SITES(CONTOURS), ERROR, and TOTAL. Each of these can be computed from the information in the analysis of variance table by dividing the sums of squares by their degrees of freedom. Variance components for CONTOURS and SITES(CONTOURS) are computed as shown earlier in this section, and the variance component for ERROR is simply MS[E].

Finally, the overall mean of the 18 measurements is -0.03834 and a measure of the precision of this estimate is its standard error, 0.15292. The *SAS/STAT User's Guide (Volume 2),* page 1132, defines this standard error thus: "Its [the mean's] variance is estimated by a certain linear combination of the estimated variance components, which is identical to the mean square due to the first factor in the model divided by the total number of observations when the design is balanced."

Figure 14.19. Analysis using the NESTED procedure

```
                    Analysis using NESTED

               Coefficients of Expected Mean Squares

          Source          CONTOURS         SITES         ERROR

          CONTOURS            6              3             1
          SITES               0              3             1
          ERROR               0              0             1

                    Analysis using NESTED

     Nested Random Effects Analysis of Variance for Variable FLUX


                 Degrees
Variance            of      Sum of                              Error
Source           Freedom    Squares    F Value    Pr > F        Term

TOTAL              17      10.421692
CONTOURS            2       0.841813     0.186     0.8392        SITES
SITES               3       6.792664     9.748     0.0015        ERROR
ERROR              12       2.787215

Variance                              Variance            Percent
Source        Mean Square            Component           of Total

TOTAL          0.613041              0.909586            100.0000
CONTOURS       0.420907             -0.307219              0.0000
SITES          2.264221              0.677318             74.4644
ERROR          0.232268              0.232268             25.5356

              Mean                           -0.03834444
              Standard error of mean          0.15291731
```

SAS provides yet another procedure, VARCOMP (variance components, which can be used to estimate variance components in random effects models. Like GLM, VARCOMP can be used to analyze either crossed or nested factors, but unlike GLM, VARCOMP assumes *random* factor levels unless otherwise specified. And like GLM, VARCOMP can handle mixed models. However, all VARCOMP does is estimate the variance components. Any *F*-tests for the significance of the factors must be per-formed by hand or using some other procedure. Refer to *SAS/STAT User's Guide (Volume 2)* for further information on the VARCOMP procedure.

14.5 Subsampling

Example 14.18. In Example 8.2, the effects of five sources of light on the four-week heights of a variety of plant were compared. In this completely randomized experiment, investigators exposed four randomly selected seedlings to each of five light sources. The study utilized the 20 seedlings as the experimental units. In Example 8.6, we used [a] one-way fixed model to analyze these data...

Now. Let's suppose that the study in Example 8.2 was actually conducted with ten pots as experimental units and two seedlings per pot. The plant height data In Table 8.2 are shown in Table 14.2 with the appropriate pot identifications.

In this experiment, there is only *one* factor being studied: type of safelight. Nevertheless, the structure of this experiment looks very much like that of the nested two-factor study described in the preceding section, and the same analyses can be used. One can act *as if* pots were a second, random effects factor, nested within levels of the first factor, which can be either a fixed or a random effects factor.

In this example, safelight is a fixed effects factor, so either the GLM or the MIXED model procedure is the appropriate one to use to analyze these data. If the types of safelight had been randomly selected from among a large population of types of safelights, then the data could be analyzed using either the GLM procedure or the NESTED procedure. Figure 14.20 shows the SAS code and Figure 14.21 shows the analysis of variance from the GLM procedure. To use the MIXED model procedure, the SAS statements are

```
proc mixed;
  class tmt pot;
  model ht = tmt;
  random pot(tmt);
```

Figure 14.20. **SAS program for subsampling with fixed treatment effects**

```
options ls=70 ps=54;
/*Table 14.2.  Subsampling with fixed
treatment effects*/
data;
   input tmt $ pot @;
     do i = 1 to 2;
        input ht @;
        output;
     end;
   drop i;
cards;
D  1 32.94 35.98
D  2 34.76 32.40
AL 1 30.55 32.64
AL 2 32.37 32.04
AH 1 31.23 31.09
AH 2 30.62 30.42
BL 1 34.41 34.88
BL 2 34.07 33.87
BH 1 35.61 35.00
BH 2 33.65 32.91
;
run;
proc glm;
   class tmt pot;
   model ht = tmt pot(tmt);
   random pot(tmt)/test;
run;
```

Recall that GLM assumes all factors are fixed effects factors unless otherwise directed. Technically speaking, the only factor here is the safelight type, and it is a fixed effects factor. However, we are analyzing the data *as if* POT(TMT) were another, random effects, factor. The table of expected mean squares in Figure 14.22 indicates that the test for treatment effects should be performed with the mean square for POT(TMT) as the denominator of the *F*-statistic. (The term "Q(TMT)" in the expected mean square for TMT simply indicates a quadratic function of the fixed treatment effects. Also, recall that it is understood that Var(Error) is the expected mean square for ERROR.) The correct test for treatment effects is shown in Figure 14.23.

Figure 14.21. GLM analysis of variance for subsampling example

```
                 General Linear Models Procedure
                    Class Level Information

            Class      Levels     Values

            TMT            5       AH AL BH BL D

            POT            2       1 2


       Number of observations in data set = 20


                      The SAS System

              General Linear Models Procedure

Dependent Variable: HT

Source              DF    Sum of Squares    F Value    Pr > F

Model                9       47.19312000      5.11      0.0089

Error               10       10.26420000

Corrected Total     19       57.45732000

           R-Square              C.V.               HT Mean

           0.821360           3.063389           33.0720000


Source              DF       Type I SS     F Value    Pr > F

TMT                  4      41.08077000     10.01      0.0016
POT(TMT)             5       6.11235000      1.19      0.3793

Source              DF      Type III SS    F Value    Pr > F

TMT                  4      41.08077000     10.01      0.0016
POT(TMT)             5       6.11235000      1.19      0.3793
```

Figure 14.22. Expected mean squares for subsampling example

```
                         General Linear Models Procedure

Source           Type III Expected Mean Square

TMT              Var(Error) + 2 Var(POT(TMT)) + Q(TMT)

POT(TMT)         Var(Error) + 2 Var(POT(TMT))
```

Figure 14.23. Test for treatment effects in the subsampling example

```
                       General Linear Models Procedure
             Tests of Hypotheses for Mixed Model Analysis of Variance

Dependent Variable: HT


Source: TMT
Error: MS(POT(TMT))
                        Denominator   Denominator
        DF    Type III MS    DF           MS     F Value  Pr > F
         4    10.2701925      5        1.22247   8.4012   0.0192

Source: POT(TMT)
Error: MS(Error)
                        Denominator   Denominator
        DF    Type III MS    DF           MS     F Value  Pr > F
         5     1.22247       10        1.02642   1.1910   0.3793
```

The p-value for the test for treatment effects, $p = 0.0192$, indicates that there is quite strong evidence that the five types of safelights differ in their effects on seedling heights. Presumably, the researcher would at this point test or compute estimates of an appropriate set of contrasts to determine which light or set of lights give optimal results. Recall that these tests or estimates should use the mean square for POT(TMT) as the "error" term.

The researcher is probably not interested in testing whether there is significant variation from pot to pot within a treatment. After all, pots are experimental units, and the reason that more than one pot is used in each treatment is because it is recognized that there is variation from one experimental unit to another. This variation is typically considered to be random, or unexplained, or due to chance.

What might be of interest, instead of a test of the significance of the POT(TMT) variation, is an estimate of its magnitude and a comparison of the size of the pot-to-pot variation to the plant-to-plant variation. The expected mean squares in Table 14.22 give

$$\hat{V}ar(ERROR) = \hat{V}ar(PLANTS) = MS[E] = 1.02$$

and

$$\hat{V}ar[POT(TMT)] = \{MS[POT(TMT)] - MS[E]\} / 2$$
$$= (1.22 - 1.02) / 2$$
$$= 0.10 \, .$$

15

Analysis of Randomized Complete Block and Latin Square Designs

15.1 Introduction

Just as single- or multiple-factor studies in a completely randomized design could be analyzed using either the ANOVA or the GLM procedure, so can single- or multiple-factor studies performed in blocks. The analysis typically involves nothing more complicated than identifying the source or sources of controlled variation in the MODEL statement.

15.2 Single-Factor Studies in Randomized Complete Blocks (RCB)

Example 15.1. A feed trial to compare three dietary supplements was conducted using 12 animals of approximately the same body weight. The 12 animals consisted of four litters, each containing three animals. Within a given litter, the three animals were randomly assigned to the three dietary supplements. The animals were housed in 12 identical pens and fed their assigned diets under identical conditions. The measured weight gains (g) after 12 weeks on the experimental diets are as follows...

The three dietary supplements are levels of a fixed effects factor, since the experiment was conducted specifically to study these three supplements. Litters is a random blocking factor, since presumably results of this study would be generalized beyond the effects of these four litters. Since there is one animal from each litter fed each of the three supplements, we will assume that no diet × litter interaction was expected to be present.

Even though there is a single factor, diet supplement, being studied here and litter is only an identifiable source of nuisance variation which we are trying to account for in order to have a more sensitive test of diet effects, the analysis looks like that of two crossed factors without interaction. Figure 15.1 shows the SAS program for analyzing these data, and Figure 15.2 shows the analysis of variance from the GLM procedure. Note that the Type III analysis in Figure 15.2 could have also been obtained by using the ANOVA procedure.

Figure 15.1. SAS program for randomized block example

```
options ls=70 ps=54;
/*Example 15.1.  Randomized blocks*/
data;
   input diet @;
      do i = 1 to 4;
         input litter gain @;
         output;
      end;
   drop i;
   lines;
1 1 28.7 2 29.3 3 28.2 4 28.6
2 1 30.7 2 34.9 3 32.6 4 34.4
3 1 31.9 2 34.2 3 34.9 4 35.3
;
run;
proc glm;
   class diet litter;
   model gain = diet litter;
   random litter;
   estimate '1vs23' diet -1 .5 .5;
   means diet;
run;
```

Figure 15.2. GLM analysis of randomized block example

```
                  General Linear Models Procedure
                     Class Level Information

                 Class     Levels     Values

                 DIET         3       1 2 3

                 LITTER       4       1 2 3 4


          Number of observations in data set = 12


             General Linear Models Procedure

Dependent Variable: GAIN

Source                DF    Sum of Squares   F Value    Pr > F

Model                  5     77.13416667      12.49     0.0040

Error                  6      7.40833333

Corrected Total       11     84.54250000

               R-Square            C.V.            GAIN Mean

               0.912371          3.475154         31.9750000


Source                DF      Type I SS     F Value    Pr > F

DIET                   2     66.06500000     26.75     0.0010
LITTER                 3     11.06916667      2.99     0.1177

Source                DF      Type III SS   F Value    Pr > F

DIET                   2     66.06500000     26.75     0.0010
LITTER                 3     11.06916667      2.99     0.1177
```

A check of the table of expected mean squares in Figure 15.3 indicates that the denominator of the F-statistic to test for differences in weight gains among diets should be MS[E]; therefore, we can conclude from the p-value, $p = 0.0010$, in Figure 15.2 that there are significant differences among the three diets.

Figure 15.3. Table of expected mean squares for fixed treatment and random block effects

```
                        General Linear Models Procedure

Source          Type III Expected Mean Square

DIET            Var(Error) + Q(DIET)

LITTER          Var(Error) + 3 Var(LITTER)
```

Figure 15.4 shows the estimate of the difference in weight gains between Diet 1 and the average of Diets 2 and 3. Since

$$t(6,0.05) = 1.943$$

the 95% lower bound for this contrast is

$$4.9125 - 1.943(0.6805)$$
$$= 3.5903.$$

Figure 15.4. Estimate of contrast between Diet 1 and Diets 2 and 3

```
                        General Linear Models Procedure

Dependent Variable: GAIN

                                T for H0:      Pr > |T|    Std Error of
Parameter           Estimate    Parameter=0                Estimate

1vs23              4.91250000        7.22      0.0004       0.68045634
```

Figure 15.5 shows means and standard deviations for weight gains on the three diets.

Figure 15.5. Means and standard deviations for weight gains on the three diets

```
                        General Linear Models Procedure

            Level of            -------------GAIN------------
            DIET        N        Mean              SD

            1           4       28.7000000       0.45460606
            2           4       33.1500000       1.90875177
            3           4       34.0750000       1.51959424
```

15.3. Randomized Complete Block Factorial Experiments

Example 15.5. The response times (msec) of eight subjects after treatment with each of four therapies intended to increase mental awareness are as follows...

In this study, the experimental design is a 2×2 factorial in randomized complete blocks. The treatments are the combinations of the levels [present or absent] of two factors: drug A and drug B.

Blocks (subjects) being an identifiable, and thus controllable, source of variation, the program shown in Figure 15.6 is simply one to analyze a 2×2 factorial experiment, with an additional term, SUBJECTS, included in the model statement. Since each $A \times B$ combination (treatment) appears eight times, it is possible to test for $A \times B$ interaction. However, since each subject receives each $A \times B$ combination *only once*, it is impossible to separate treatment $\times$ block interaction from error.

The GLM analysis is shown in Figure 15.7, and the table of expected mean squares, assuming that both factors are fixed effects factors and that subjects (blocks) are random effects, is shown in Figure 15.8. The expected mean squares indicate that MS[E] is the appropriate denominator for the F-statistics to test both main effects and the interaction.

Figure 15.6. SAS program to analyze 2 × 2 factorial in eight blocks

```
options ls=70 ps=54;
data a;
input A$ B$ @;
  do i = 1 to 8;
    input subject time @;
    output;
  end;
  drop i;
lines;
 no  no 1 18.8 2 18.5 3 21.4 4 25.5 5 19.8 6
24.4 7 25.6 8 26.5
yes  no 1 13.5 2  9.8 3 12.1 4 22.9 5  8.3 6
14.9 7 23.0 8 15.5
 no yes 1 13.6 2 13.4 3  8.3 4 24.9 5 16.9 6
16.2 7 16.3 8 15.4
yes yes 1 10.6 2 12.6 3 11.7 4 16.8 5  4.0 6
13.4 7 14.9 8 12.6
;
run;
proc glm;
  class A B subject;
  model time = subject A B A*B/p;
  random subject;
  means A B A*B;
  output out = b p=ybar1;
run;
proc sort data = b;
  by A B;
run;
proc means data = b;
  var ybar1;
  by A B;
  output out = means mean = ybar;
run;
proc plot data = means;
  plot ybar*B = A;
run;
```

Figure 15.7. GLM analysis for 2 x 2 factorial in 8 blocks

```
                    General Linear Models Procedure
                        Class Level Information

            Class     Levels    Values

            A            2      no yes

            B            2      no yes

            SUBJECT      8      1 2 3 4 5 6 7 8


            Number of observations in data set = 32
```

Dependent Variable: TIME

Source	DF	Sum of Squares	F Value	Pr > F
Model	10	839.63812500	10.31	0.0001
Error	21	171.04406250		
Corrected Total	31	1010.68218750		

R-Square	C.V.	TIME Mean
0.830764	17.49205	16.3156250

Source	DF	Type I SS	F Value	Pr > F
SUBJECT	7	365.92468750	6.42	0.0004
A	1	246.97531250	30.32	0.0001
B	1	194.53781250	23.88	0.0001
A*B	1	32.20031250	3.95	0.0600

Source	DF	Type III SS	F Value	Pr > F
SUBJECT	7	365.92468750	6.42	0.0004
A	1	246.97531250	30.32	0.0001
B	1	194.53781250	23.88	0.0001
A*B	1	32.20031250	3.95	0.0600

Figure 15.8. Expected mean squares for two fixed effects factors in random blocks

```
              General Linear Models Procedure
Source          Type III Expected Mean Square

SUBJECT         Var(Error) + 4 Var(SUBJECT)

A               Var(Error) + Q(A,A*B)

B               Var(Error) + Q(B,A*B)

A*B             Var(Error) + Q(A*B)
```

As was the case with the single-factor experiment performed in blocks, removing the block effect from error in this two-factor study decreases the error sum of squares and thus provides for a more sensitive test of interaction and main effects. No significant interaction between factors *A* and *B* is detected, but both main effects are found to be highly significant. The table of means and standard deviations in Figure 15.9 and the plot in Figure 15.10 may be examined to give some indication of which drug combinations are optimal.

Figure 15.9. Row, column, and cell means for response times

```
              General Linear Models Procedure

    Level of          -------------TIME------------
    A          N         Mean              SD

    no         16      19.0937500        5.29206560
    yes        16      13.5375000        4.78621284

    Level of          -------------TIME------------
    B          N         Mean              SD

    no         16      18.7812500        5.85607590
    yes        16      13.8500000        4.48508640

Level of  Level of      -------------TIME------------
A         B        N        Mean              SD

no        no       8      22.5625000        3.30321813
no        yes      8      15.6250000        4.64750317
yes       no       8      15.0000000        5.46756671
yes       yes      8      12.0750000        3.78219513
```

Figure 15.10. Plots of treatment means for mental awareness experiment

```
                      Plot of YBAR*B.   Symbol is value of A.

          24 +
             |
             |
             |
             |
             |   n
             |
          22 +
             |
             |
             |
             |
             |
          20 +
             |
             |
      YBAR   |
             |
             |
          18 +
             |
             |
             |
             |
             |
          16 +
             |
             |                                                  n
             |   y
             |
             |
          14 +
             |
             |
             |
             |
             |
          12 +                                                  y
             ---+----------------------------------------------+--
               no                                             yes
                                    B
```

15.4 The Latin Square Design

Example 15.8. Researchers decide to compare four methods of determining impurities in a specimen sample. Because a measured impurity value can be affected both by the analyst who analyzes the specimen and by the batch of the raw material used to prepare the specimen, the experiment was based on a Latin Square design with rows representing analysts and columns representing the batch of the raw material. Four analysts (who could be considered a random sample from a population of analysts) and a random sample of four batches of raw material were used in the study.

The analysis of this Latin Square design is very similar to the analysis of the single-factor RCB in Section 15.2, with simply the addition of another source of identifiable variation in the MODEL statement. Since each method is analyzed only once by each analyst, the analyst × method

Figure 15.11. SAS program for a single-factor Latin Square design

```
options ls=70 ps=54;
/*Example 15.8.  Single-factor Latin Square*/
data;
  input analyst @;
    do i = 1 to 4;
      input batch pct method @;
      output;
    end;
  drop i;
lines;
1 1 3.10 2 2 4.29 3 3 6.53 4 4 3.52 1
2 1 4.21 3 2 8.30 4 3 3.92 1 4 4.82 2
3 1 6.11 1 2 2.98 2 3 4.20 3 4 8.46 4
4 1 9.01 4 2 4.59 1 3 4.60 2 4 5.22 3
;
run;
proc glm;
  class analyst batch method;
  model pct = analyst batch method;
  random analyst batch;
run;
```

interaction is indistinguishable from error. Similarly, since each method is used only once on each batch, we cannot model batch × method interaction, and since each analyst analyzes material from each batch only once, we cannot model analyst × batch interaction. Thus, the SAS program in Figure 15.11 for analyzing this single-factor Latin Square looks the same as a program to analyze a three-factor main-effects only (no interaction) model. Keep in mind, however, that the method of assigning experimental units to treatments makes the two designs quite different, even though the analyses look the same.

The GLM analysis of variance is shown in Figure 15.12. The table of expected mean squares for both blocking variables random effects and the experimental factor fixed effects in Figure 15.13 indicates that the denominator of the F-statistic for testing for differences among the four methods should be MS[E] so that the default analysis is appropriate. Figure 15.12 indicates that there are significant ($p = 0.0016$) differences among the four methods. All pairwise comparisons or some *a priori* contrasts would be tested at this point to determine which methods differ from which others, and how.

Figure 15.12. GLM analysis for Latin Square design

```
                    General Linear Models Procedure
                       Class Level Information

               Class      Levels     Values

               ANALYST       4       1 2 3 4

               BATCH         4       1 2 3 4

               METHOD        4       1 2 3 4


          Number of observations in data set = 16


                    General Linear Models Procedure

Dependent Variable: PCT

Source              DF     Sum of Squares     F Value     Pr > F

Model                9        50.39992500        7.62     0.0112

Error                6         4.40785000

Corrected Total     15        54.80777500

               R-Square              C.V.              PCT Mean

               0.919576            16.35321          5.24125000


Source              DF       Type I SS      F Value      Pr > F

ANALYST              3       4.78752500        2.17      0.1923
BATCH                3       1.71212500        0.78      0.5482
METHOD               3      43.90027500       19.92      0.0016

Source              DF      Type III SS      F Value      Pr > F

ANALYST              3       4.78752500        2.17      0.1923
BATCH                3       1.71212500        0.78      0.5482
METHOD               3      43.90027500       19.92      0.0016
```

Figure 15.13. Expected mean squares for Latin Square design (fixed effects experimental factor, random effects blocking factors)

```
                    General Linear Models Procedure

Source          Type III Expected Mean Square

ANALYST         Var(Error) + 4 Var(ANALYST)

BATCH           Var(Error) + 4 Var(BATCH)

METHOD          Var(Error) + Q(METHOD)
```

15.5 Checking the Validity of Assumptions

In order to check the additivity assumption for Example 15.1, we want to plot residuals against predicted values. The program in Figure 15.1 can be modified to produce the desired graph by inserting the command

```
output out=check p=ybar r=r;
```

after the model statement and adding

```
proc plot data=check;
   plot r*ybar;
   run;
```

at the end of the program. No definite pattern indicating lack of additivity can be seen in the plot so produced (see Figure 12.14).

The UNIVARIATE procedure can be used to check whether the residuals show significant departure from normality. The code necessary to produce Figure 15.15 is

```
proc univariate normal plot data=check;
            var r;
```

Note that only those portions of the UNIVARIATE procedure pertinent to the normality check are shown in Figure 15.15.

Figure 15.14. Residual plot for RCB example

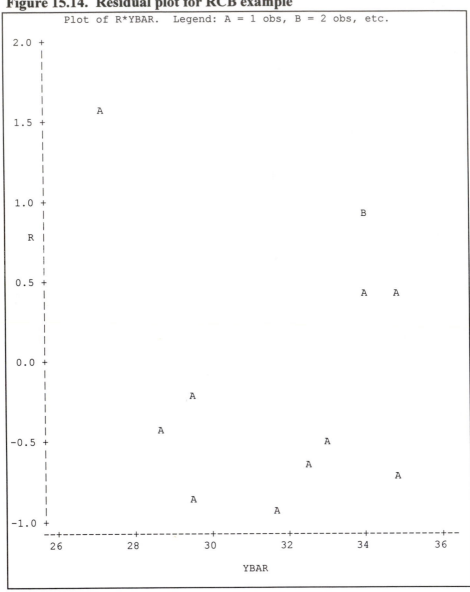

Figure 15.15. **Checking the assumption of normally distributed residuals**

```
                            Univariate Procedure

Variable=R

                               Moments

             N                12    Sum Wgts          12
             Mean              0    Sum                0
             Std Dev    0.820661    Variance    0.673485
             Skewness   0.614139    Kurtosis   -0.93574
             USS        7.408333    CSS         7.408333
             CV                .    Std Mean    0.236904
             T:Mean=0          0    Pr>|T|        1.0000
             Num ^= 0         12    Num > 0            5
             M(Sign)          -1    Pr>=|M|       0.7744
             Sgn Rank          0    Pr>=|S|       1.0000
             W:Normal   0.904662    Pr<W          0.1738


      Stem Leaf                        #          Boxplot
         1 5                           1             |
         1                                           |
         0 599                         3          +-----+
         0 4                           1          |  +  |
        -0 42                          2          *-----*
        -0 99765                       5          +-----+
           ----+----+----+----+




                            Univariate Procedure

Variable=R

                         Normal Probability Plot
       1.75+                                     *   ++++++
           |                                  ++++++
           |                               +*++*+
           |                           +++*+*+
           |                       +++*++* *
      -0.75+           *     +*++*+ *
           +----+----+----+----+----+----+----+----+----+----+
             -2        -1         0        +1        +2
```

16

Repeated Measures Studies

16.1 Introduction

Until now (with the exception of the paired *t*-test in Chapter 4 and subsampling in Chapter 14), the analyses we have performed have assumed that subjects (or cases, or observational units) were nested within treatments. The SAS procedures we have used so far generally assume nested cases by default, that is, it is not necessary to specify that subjects/cases are nested within treatments in procedures such as ANOVA or GLM.

In repeated measures studies, every subject/case/observational unit either gets every treatment (combination of factor levels), or gets every level of at least one of the factors. This was the case in the paired *t*-test in situations in which a single item was measured under two conditions, such as before and after a treatment was applied. A convenient way of thinking about a repeated measures design is that subjects/cases are *crossed* with the levels of one or more factors. The trick to get SAS to analyze repeated measures studies, then, is to model subjects as if they were a random effects factor and to use the appropriate subject-by-factor interaction as the error term for testing that factor. This works for both single-factor and multiple-factor repeated measures studies.

A variation on the theme of the classical repeated measures design is the split-plot design, in which the effects of two or more factors, and their interaction(s), are being studied. In split-plot designs, practical considerations dictate that levels of one of the factors (call it *A*) must be applied to a whole group of experimental units at once. Once that is done, levels of the other factor (*B*) are randomly assigned to experimental units which have already had a level of the first factor (A) applied. Thus, if the group of experimental units receiving the same level of A are considered to be a "subject", then there are repeated measures on factor B: every "subject" receives every level of factor B. (Note the analogy with the subsampling scheme discussed in Rao's Section 14.6, in which experimental units and observational units differ.)

16.2 Split-Plot Experiments

Example 16.5. In an experiment three formulations of a diet (factor A) were compared on the basis of the amounts of a particular chemical in the diet that were absorbed by the kidneys of experimental rats. The investigators were also interested in comparing three techniques (factor B) for measuring the absorbed amounts. Four litters, each containing three rats, were used in the study. Within each litter, the animals were randomly assigned to the three formulations. After two weeks on the diets, the animals were sacrificed, and three sample specimens were selected from each animal's kidney. The three methods were randomly assigned to the three specimens, and the absorbed amounts were measured for the specimens using the assigned methods. The data (percentage of chemical absorbed) are given in Table 16.3.

The study design is a split-plot deign in randomized complete blocks. Each of the $r = 4$ blocks (litters) contains $a = 3$ main-plots (individual rats) and each main-plot contains $b = 3$ sub-plots (specimens). The 3 formulations are the main-plot treatments and the 3 methods are the sub-plot treatments.

If absorption amounts in the three specimens (sub-plots) from a single rat (main-plot) are uncorrelated, then the data can be analyzed as data from a 3×3 factorial experiment in a randomized complete block design. If, on the other hand, it is suspected that correlation might exist between measurements from the specimens of the same rat (main-plot), then a model that allows for correlations between repeated measures should be used for analyzing the absorption data.

Example 16.6. The design consists of main plot and subplot components. In the RCB main plot design, three formulations (levels of A) are the treatments, the litters are the blocks, and the rats (main units) are the experimental units...

The subplot design is composed of $a = 3$ RCB designs, one for each formulation. The main units are the blocks and the sub-treatments are the treatments in the subplot designs.

Since the data set provided in Rao's Example 16.5 is a little bit complex it was entered in a SAS file called SPLTPLT.DAT. This data file is shown in Figure 16.1.   The program to read and analyze the data as a split-plot experiment is given in Figure 16.2.

Figure 16.1. Data file for split-plot experiment

```
1 1 1  26.97 2 26.12 3 27.83 4 27.47
  2 1  22.60 2 22.91 3 19.83 4 21.63
  3 1  30.71 2 29.53 3 27.51 4 28.62
2 1 1  17.47 2 18.13 3 18.01 4 17.97
  2 1  16.90 2 16.31 3 16.52 4 15.93
  3 1  23.95 2 22.84 3 23.84 4 23.45
3 1 1  20.72 2 20.41 3 21.01 4 21.34
  2 1  24.32 2 25.06 3 25.92 4 25.33
  3 1  28.31 2 29.02 3 29.13 4 29.36
```

Refer to Rao's Box 16.4 in which the following are defined:

μ = the overall mean

α_i = the fixed effect of the i-th level of A

β_j = the fixed effect of the j-th level of B

$(\alpha\beta)_{ij}$ = the joint effect of the i-th level of A and the j-th level of B

R_k = the random effect of the k-th block

Figure 16.2. SAS Program for split-plot experiment

```
options ls=70 ps=54;
data a;
  infile 'spltplt.dat';
  do k = 1 to 3;
  input form @@;
    do i = 1 to 3;
    input method @@;
      do j = 1 to 4;
      input litter pct @@;
      output;
      end;
    end;
  end;
drop i; drop j; drop k;
run;
proc glm;
  class litter form method;
  model pct = litter form litter*form
                method form*method;
  test h=form e=litter*form;
  estimate '22vs23' method 0 1 -1 form*method
      0 0 0 0 1 -1 0 0 0;
  means form/lsd e=litter*form;
  means method/lsd;
run;
```

$(SR)_{ik}$ = the joint effect of the i-th level of A and the k-th block, which can also be denoted as $(AR)_{ik}$

and

E_{ijk} = the error effect.

If we were to treat this as a three formulations × three methods factorial experiment in four blocks, we would write the model as

$$Y_{ijk} = \mu + \alpha_i + \beta_j + (\alpha\beta)_{ij} + R_k + (\alpha R)_{ik} + (\beta R)_{jk} + (\alpha\beta R)_{ijk}$$

where the last term would typically be written as E_{ijk}.

However, as pointed out in Rao's Example 16.6, the split-plot design is really two designs: a randomized complete blocks design for factor A with litters (R) as blocks, and then another three randomized complete blocks designs with levels of factor A now serving as blocks and factor B being the treatment. So let us rearrange the terms of the above model to reflect this:

$$Y_{ijk} = \mu + [R_k + \alpha_i + (\alpha R)_{ik}] + \beta_j + (\alpha\beta)_{ij} + (\beta R)_{jk} + (\alpha\beta R)_{ijk}$$

$$= \mu + [R_k + \alpha_i + (\alpha R)_{ik}] + \beta_j + (\alpha\beta)_{ij} + E_{ijk}.$$

The factor A-by-block interaction mean square is used as the error term to test for significant differences in the responses under different levels of factor A; any effects involving factor B are tested with the error mean square. Leaving the terms

$$(\beta R)_{jk} + (\alpha\beta R)_{ijk}$$

out of the MODEL statement in GLM will cause SAS to define variation due to them as "error variation." Including the $A \times$ block interaction in the MODEL statement allows us to specify that its mean square is to be used as the error term for testing factor A effects.

Output from the program is shown in Figures 16.3-16.6. (Since the design is balanced, the Type I analysis has been deleted from Figure 16.3.) The Type III F-tests that appear by default in the analysis of variance table are computed using the error mean square as the denominator; the proper Type III F-value for testing FORM effects is obtained by using the TEST statement in GLM and is shown at the bottom of Figure 16.3. Note that if a test on LITTER effects were desired, the proper denominator would be the LITTER*FORM mean square, resulting in an F-value of $F = 0.11$ and $p = 0.9520$, as in Rao's Example 16.7.

The presence of a significant FORM*METHOD interaction indicates that comparisons among methods must be made for each formulation separately, and vice versa. The ESTIMATE statement in the program in Figure 16.2 compares Methods 2 and 3 under Formulation 2; its output is shown in Figure 16.4. Since

Figure 16.3. Analysis of data from split-plot experiment: class levels information and analysis of variance

```
                 General Linear Models Procedure
                    Class Level Information

              Class      Levels     Values

              LITTER        4       1 2 3 4

              FORM          3       1 2 3

              METHOD        3       1 2 3


          Number of observations in data set = 36
```

Dependent Variable: PCT

Source	DF	Sum of Squares	F Value	Pr > F
Model	17	679.71750556	70.14	0.0001
Error	18	10.26055000		
Corrected Total	35	689.97805556		

R-Square	C.V.	PCT Mean
0.985129	3.224291	23.4161111

Source	DF	Type III SS	F Value	Pr > F
LITTER	3	0.34014444	0.20	0.8958
FORM	2	314.23167222	275.63	0.0001
LITTER*FORM	6	6.26370556	1.83	0.1493
METHOD	2	260.57357222	228.56	0.0001
FORM*METHOD	4	98.30841111	43.12	0.0001

Tests of Hypotheses using the Type III MS for
LITTER*FORM as an error term

Source	DF	Type III SS	F Value	Pr > F
FORM	2	314.23167222	150.50	0.0001

$$t(18, 0.025) = 2.101,$$

we have a two-sided 95% confidence interval on the difference in the mean responses that is, as in Rao's Example 16.7 with the exception of the direction of the difference,

Figure 16.4. Estimate of the difference between Methods 2 and 3 under Formulation 2

```
                       General Linear Models Procedure

Dependent Variable: PCT

                                 T for H0:     Pr > |T|    Std Error of
Parameter              Estimate  Parameter=0               Estimate

22vs23              -7.10500000       -13.31     0.0001    0.53386822
```

$$-7.105 \pm 2.101(0.5339)$$
$$= -7.105 \pm 1.122.$$

Note, however, that the ESTIMATE command in GLM cannot be made to compute the standard error of the estimate using any other mean square than the error mean square. Therefore, had it been desired to compare formulations under a given method, you would have had to compute the appropriate standard error by hand. The CONTRAST statement can be made to test for the significance of the contrast using the appropriate standard error, however, by specifying

```
/e=litter*form
```

in the CONTRAST statement.

Had no significant interaction between methods and formulations been present, it would have been appropriate to perform pairwise comparisons among formulation means and among methods means. Figure 16.5 shows pairwise comparisons among the factor A levels. Note that use of the formulation $\times$ litters mean square is specified as the error term for comparing these means. Figure 16.6 gives pairwise comparisons among methods, using the error mean square as the error term.

Figure 16.5. Pairwise comparisons among factor *A* levels

```
                 General Linear Models Procedure

                   T tests (LSD) for variable: PCT

     NOTE: This test controls the type I comparisonwise error rate
           not the experimentwise error rate.

               Alpha= 0.05  df= 6  MSE= 1.043951
                     Critical Value of T= 2.45
               Least Significant Difference= 1.0207

     Means with the same letter are not significantly different.

            T Grouping           Mean      N  FORM

                    A          25.9775    12  1
                    A
                    A          24.9942    12  3

                    B          19.2767    12  2
```

Figure 16.6. Pairwise comparisons among methods

```
                 General Linear Models Procedure

                   T tests (LSD) for variable: PCT

     NOTE: This test controls the type I comparisonwise error rate
           not the experimentwise error rate.

               Alpha= 0.05  df= 18  MSE= 0.570031
                     Critical Value of T= 2.10
               Least Significant Difference= 0.6476

     Means with the same letter are not significantly different.

            T Grouping           Mean      N  METHOD

                    A          27.1892    12  3

                    B          21.9542    12  1

                    C          21.1050    12  2
```

16.3 Repeated Measures: Univariate Approach

In order to get a complete analysis of repeated measures data using SAS, it is often necessary to analyze the data two different ways: the first way is pretty much the same as was used for the split-plot data, and the second is by using the REPEATED statement in the GLM procedure. In this section we will show how to analyze the data using a univariate approach. The multivariate approach using the REPEATED command will be taken up in Section 16.4.

> **Example 16.2.** Potoff and Roy (1964) reported data on dental measurements (distances, mm, from the center of the pituitary to the maxillary fissure) of $n = 27$ children (11 girls and 16 boys) at the University of North Carolina Dental School. Each child was measured four times over a period of six years--at 8, 10, 12, and 14 years of age.
>
> How can we analyze these data to evaluate differences over time in the dental measurements of male and female children? One possibility is to use the general linear model in which gender (A) and age (B) are explanatory factors...
>
> ...The assumption that the errors are independent is unlikely to be satisfied by the dental data, because a measurement of a child at a given age will most likely be correlated with repeated measurements of the same child at other ages.

Figure 16.7 shows the Potoff and Roy data as presented in Rao's Example 16.8. Because this is a rather large and complex data set, it was entered into a SAS data file named DENTAL.DAT. Some careful planning should go into the data entry for repeated measures studies, because the two different approaches for analysis--univariate and multivariate--require the data to be entered in different ways. It is helpful if the data have to be typed in only once, in a format that can be used by both approaches. Why this is necessary will not be apparent until after both approaches have been illustrated. It is helpful at this point to examine how the dental measurements data were entered into the DENTAL.DAT file.

Figure 16.7. Dental measurements data

```
girl  1 1 21.0 2 20.0 3 21.5 4 23.0
      2 1 21.0 2 21.5 3 24.0 4 25.5
      3 1 20.5 2 24.0 3 24.5 4 26.0
      4 1 23.5 2 24.5 3 25.0 4 26.5
      5 1 21.5 2 23.0 3 22.5 4 23.5
      6 1 20.0 2 21.0 3 21.0 4 22.5
      7 1 21.5 2 22.5 3 23.0 4 25.0
      8 1 23.0 2 23.0 3 23.5 4 24.0
      9 1 20.0 2 21.0 3 22.0 4 21.5
     10 1 16.5 2 19.0 3 19.0 4 19.5
     11 1 24.5 2 25.0 3 28.0 4 28.0
boy  12 1 26.0 2 25.0 3 29.0 4 31.0
     13 1 21.5 2 22.5 3 23.0 4 26.5
     14 1 23.0 2 22.5 3 24.0 4 27.5
     15 1 25.5 2 27.5 3 26.5 4 27.0
     16 1 20.0 2 23.5 3 22.5 4 26.0
     17 1 24.5 2 25.5 3 27.0 4 28.5
     18 1 22.0 2 22.0 3 24.5 4 26.5
     19 1 24.0 2 21.5 3 24.5 4 25.5
     20 1 23.0 2 20.5 3 31.0 4 26.0
     21 1 27.5 2 28.0 3 31.0 4 31.5
     22 1 23.0 2 23.0 3 23.5 4 25.0
     23 1 21.5 2 23.5 3 24.0 4 28.0
     24 1 17.0 2 24.5 3 26.0 4 29.5
     25 1 22.5 2 25.5 3 25.5 4 26.0
     26 1 23.0 2 24.5 3 26.0 4 30.0
     27 1 22.0 2 21.5 3 23.5 4 25.0
```

First, note that this study can be characterized as a one between-subjects (SEX) and one within-subjects (AGE) repeated measures design. The design is also unbalanced, since there are more boys than girls in the study; however, fortunately, there are no missing data on any subject. In Figure 16.7 we see that the first level of the between-subjects variable is entered, followed by the subject number, with each subject being assigned a *unique* subject number. Then four pairs of measurements follow, with the first member of the pair being the age period and the second member being the measurement made at that age.

Figure 16.8 shows a SAS program to read and print the data and to graph mean measurements across time separately for girls and boys.

Figure 16.8. SAS program to read dental measurements data and construct a plot

```
options ls=70 ps=54;
data long;
  infile 'dental.dat';
  input sex $ @@;
    if sex = 'girl' then s = 11;
    if sex = 'boy' then s = 16;
    do i = 1 to s;
      input subject @@;
      do j = 1 to 4;
        input age mm @@;
        output;
      end;
    end;
  drop i; drop j;
run;
proc print data=long;
run;
proc sort; by sex age;
run;
proc means; var mm; by sex age;
  output out=means mean=ybar; run;
proc plot data=means;
  plot ybar*age=sex;
run;
```

A portion of the output from the PRINT procedure is shown in Figure 16.9. The data are called "long" in the program in Figure 16.8 because the printout in Figure 16.9 lists 108 data lines. A case is a *subject at an age*, and a single measurement is taken on each subject at each age. (We will see in the next section, why it is necessary to enter the data in such a way so that each *individual child* is a single case, with multiple measurements taken on each child, across time.)

Figure 16.9. Dental data read in the "long" format

OBS	SEX	S	SUBJECT	AGE	MM
1	girl	11	1	1	21.0
2	girl	11	1	2	20.0
3	girl	11	1	3	21.5
4	girl	11	1	4	23.0
5	girl	11	2	1	21.0
6	girl	11	2	2	21.5
7	girl	11	2	3	24.0
8	girl	11	2	4	25.5
9	girl	11	3	1	20.5
10	girl	11	3	2	24.0
11	girl	11	3	3	24.5
12	girl	11	3	4	26.0
.	.	.	.	.	.
.	.	.	.	.	.
.	.	.	.	.	.
101	boy	16	26	1	23.0
102	boy	16	26	2	24.5
103	boy	16	26	3	26.0
104	boy	16	26	4	30.0
105	boy	16	27	1	22.0
106	boy	16	27	2	21.5
107	boy	16	27	3	23.5
108	boy	16	27	4	25.0

In order to obtain a graph showing effects of sex, age, and their interaction, the data are sorted by sex and then age, and averaged across subjects. Mean measurements are outputted to a data set and plotted against age, with different symbols for data from boys and girls. Output from the MEANS procedure is not shown; the graph is shown in Figure 16.10. There are apparent sex effects, with boys having larger values than girls at all ages. There are also apparent age effects, with the measurements increasing over time for both boys and girls, on the average. There is possibly also an interaction between age and sex, as the rate of increase of the measurements with age appears to be slightly higher for boys than for girls. To test for the significance of these effects, examine the program in Figure 16.11. (Note that the data step, which is the same as in Figure 16.8, has been omitted from Figure 16.11.)

Figure 16.10. Graph of mean measurements across age, by sex

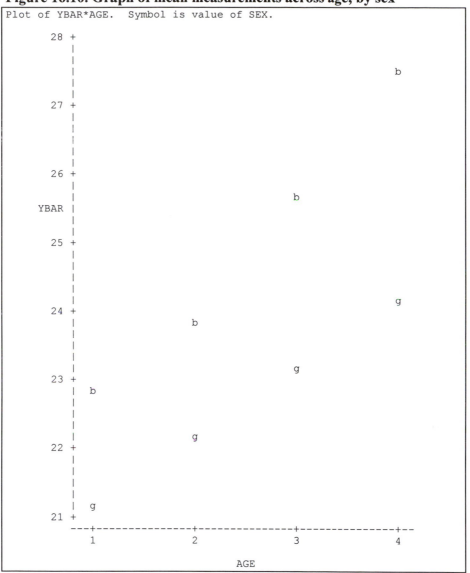

```
Plot of YBAR*AGE.   Symbol is value of SEX.

    28 +
       |
       |
       |                                                      b
       |
       |
    27 +
       |
       |
       |
       |
    26 +
       |
       |                                       b
  YBAR |
       |
    25 +
       |
       |
       |
       |                                                         g
    24 +
       |                         b
       |
       |
       |                                       g
    23 +
       |   b
       |
       |
       |                      g
    22 +
       |
       |
       |
       | g
    21 +
       ---+---------------+---------------+---------------+--
          1               2               3               4
                               AGE
```

Figure 16.11. Univariate approach to analyzing repeated measures data

```
proc glm data=long;
  class sex subject age;
  model mm =sex age sex*age subject(sex);
  test h=sex e=subject(sex);
  lsmeans sex/e=subject(sex) stderr pdiff;
  means age/lsd;
run;
```

Recall from Section 16.1 that subjects are *nested* within levels of the between-subjects variables but *crossed* with levels of the within-subjects variables in a repeated measures study. Treating SUBJECT as if it were a random effects factor and following the convention explained in Section 13.6 for writing model statements with nested factors, the MODEL statement reads

```
model mm = sex age sex*age subject(sex);
```

with the

```
subject(sex)*age
```

interaction omitted from the MODEL statement to serve as an error term. The significance of the SEX factor is tested using the SUBJECT(SEX) mean square as an error term. Analysis produced by this program is shown in Figure 16.12. Since the design is unbalanced, the Type I and Type III analyses are not identical; however, the Type I analysis has been omitted from Figure 16.12.

The slight interaction between sex and age that is seen in the plot in Figure 16.10 is shown in Figure 16.12 to be marginally significant ($p = 0.0781$). Whether or not there is a faster rate of increase in the measurement with age for boys than for girls needs to be decided by the dental professionals whose data this is. Even in the presence of an interaction,

Figure 16.12. Univariate approach to analysis of one between-subjects and one within-subjects repeated measures

```
                    General Linear Models Procedure
                       Class Level Information

Class      Levels    Values

SEX           2       boy girl

SUBJECT      27       1 2 3 4 5 6 7 8 9 10 11 12 13 14 15 16 17 18 19
                      20 21 22 23 24 25 26 27

AGE           4       1 2 3 4


                 Number of observations in data set = 108


Dependent Variable: MM

Source                  DF    Sum of Squares    F Value    Pr > F

Model                   32      769.56428872     12.18     0.0001

Error                   75      148.12784091

Corrected Total        107      917.69212963

                  R-Square            C.V.               MM Mean

                  0.838587          5.850026           24.0231481


Source                  DF      Type III SS    F Value    Pr > F

SEX                      1      140.46485690     71.12     0.0001
AGE                      3      209.43697391     35.35     0.0001
SEX*AGE                  3       13.99252946      2.36     0.0781
SUBJECT(SEX)            25      377.91477273      7.65     0.0001

Tests of Hypotheses using the Type III MS for
SUBJECT(SEX) as an error term

Source                  DF      Type III SS    F Value    Pr > F

SEX                      1      140.46485690      9.29     0.0054
```

however, there is strong evidence of the sex difference ($p = 0.0054$) and of differences over time ($p = 0.0001$). Should comparisons be made between the sexes (although this is somewhat superfluous here, there being only two sexes), they should be made using the SUBJECT(SEX) mean square as the error term, while comparisons from one age group to another should be made with reference to the SUBJECT(SEX)*AGE = ERROR error term. Note also that since the design is unbalanced with respect to sex, the MEANS statement should not be used to compare the average for males to the average for females, as these averages would not be adjusted for imbalance. Figure 16.13 shows the LSMEANS analysis for SEX and the MEANS comparison for AGE. Had other contrasts between age level means been desired, they could have been computed and tested in the usual fashion.

Figure 16.13. Pairwise comparisons for SEX and AGE levels

```
                  General Linear Models Procedure
                      Least Squares Means

     Standard Errors and Probabilities calculated using the Type III MS
                   for SUBJECT(SEX) as an Error term

      SEX            MM        Std Err    Pr > |T|      Pr > |T| H0:
                  LSMEAN        LSMEAN   H0:LSMEAN=0   LSMEAN1=LSMEAN2

      boy       24.9687500    0.4860008     0.0001        0.0054
      girl      22.6477273    0.5861390     0.0001


                  T tests (LSD) for variable: MM

        NOTE: This test controls the type I comparisonwise error rate
              not the experimentwise error rate.

              Alpha= 0.05  df= 75  MSE= 1.975038
                   Critical Value of T= 1.99
              Least Significant Difference= 0.762

        Means with the same letter are not significantly different.

              T Grouping           Mean      N  AGE

                     A           26.0926     27   4

                     B           24.6481     27   3

                     C           23.1667     27   2

                     D           22.1852     27   1
```

16.4 Repeated Measures: Multivariate Approach

The analysis of the dental measurements data in Section 16.3 may or may not be sufficient. That analysis assumes that the sphericity assumption, described in Section 16.4 of Rao, is tenable. However, when repeated measurements are taken over time, there is always the suspicion that sets of measurements taken close together in time might be more highly correlated than those more widely separated in time. In such situations, it is prudent to perform a modified F-test on the within-subjects variable. SAS can be made to perform both the Greenhouse and Geisser adjustment and the Huynh and Feldt adjustment to the standard F-test, if the REPEATED statement in GLM is used.

Refer back to Figure 16.9, in which the dental measurements data are read in the "long" format. Recall that in Figure 16.9, a case is *a child at an age*, and that a single measurement is made on each child at each age. This measurement is the single dependent variable in the analysis in Section 16.3.

Another way of thinking about the dental measurements data, however, is that each child is a case, and that there are multiple (repeated) measurements made on each case or subject. When there are multiple dependent variables, we say that we have a *multivariate* data structure. The GLM procedure in SAS can be used to model multiple dependent variables.

First, the data in Figure 16.7 need to be read in such a way that a child is a case. The program in Figure 16.14 accomplishes this, as can be seen by the printout from the PRINT procedure in Figure 16.15. Since there are 27 lines and 12 columns in Figure 16.15, as contrasted with 108 lines and six columns in the "long" format in Figure 16.9, we refer to the layout in Figure 16.15 as the "wide" format. Often, repeated measures data need to be analyzed from both the univariate standpoint (to obtain interaction plots, for example) and from the multivariate perspective (to obtain adjustments for

Figure 16.14. Program to read dental measurements data in the "wide" format

```
options ls=70 ps=54;
data wide;
  infile 'dental.dat';
  input sex $ @@;
  if sex = 'girl' then s = 11;
  if sex = 'boy' then s = 16;
    do i = 1 to s;
    input subject @@;
    input age1 mm1 age2 mm2 age3 mm3 age4
      mm4;
    output;
    end;
  drop i;
run;
proc print data=wide;
run;
```

Figure 16.15. Dental measurements data in the "wide" format

OBS	SEX	S	SUBJECT	AGE1	MM1	AGE2	MM2	AGE3	MM3	AGE4	MM4
1	girl	11	1	1	21.0	2	20.0	3	21.5	4	23.0
2	girl	11	2	1	21.0	2	21.5	3	24.0	4	25.5
3	girl	11	3	1	20.5	2	24.0	3	24.5	4	26.0
4	girl	11	4	1	23.5	2	24.5	3	25.0	4	26.5
5	girl	11	5	1	21.5	2	23.0	3	22.5	4	23.5
6	girl	11	6	1	20.0	2	21.0	3	21.0	4	22.5
7	girl	11	7	1	21.5	2	22.5	3	23.0	4	25.0
8	girl	11	8	1	23.0	2	23.0	3	23.5	4	24.0
9	girl	11	9	1	20.0	2	21.0	3	22.0	4	21.5
10	girl	11	10	1	16.5	2	19.0	3	19.0	4	19.5
11	girl	11	11	1	24.5	2	25.0	3	28.0	4	28.0
12	boy	16	12	1	26.0	2	25.0	3	29.0	4	31.0
13	boy	16	13	1	21.5	2	22.5	3	23.0	4	26.5
14	boy	16	14	1	23.0	2	22.5	3	24.0	4	27.5
15	boy	16	15	1	25.5	2	27.5	3	26.5	4	27.0
16	boy	16	16	1	20.0	2	23.5	3	22.5	4	26.0
17	boy	16	17	1	24.5	2	25.5	3	27.0	4	28.5
18	boy	16	18	1	22.0	2	22.0	3	24.5	4	26.5
19	boy	16	19	1	24.0	2	21.5	3	24.5	4	25.5
20	boy	16	20	1	23.0	2	20.5	3	31.0	4	26.0
21	boy	16	21	1	27.5	2	28.0	3	31.0	4	31.5
22	boy	16	22	1	23.0	2	23.0	3	23.5	4	25.0
23	boy	16	23	1	21.5	2	23.5	3	24.0	4	28.0
24	boy	16	24	1	17.0	2	24.5	3	26.0	4	29.5
25	boy	16	25	1	22.5	2	25.5	3	25.5	4	26.0
26	boy	16	26	1	23.0	2	24.5	3	26.0	4	30.0
27	boy	16	27	1	22.0	2	21.5	3	23.5	4	25.0

departures from sphericity), so some care needs to be taken in data entry so that a single data file can be read in both the "long" and the "wide" formats.

In Figure 16.15, four dependent variables, MM1, MM2, MM3, and MM4 are made on each child. The program in Figure 16.16 shows these four variables all listed on the left side of the model statement in the GLM procedure. (The data step in Figure 16.16 is the same as in Figure 16.14.) The between-subjects variable, SEX, is listed on the right side of the MODEL statement. The option

```
/nouni;
```

stands for "no univariate analyses". If this option is not requested, then the difference between boys' and girls' measurements will be analyzed separately at each of the four time periods.

Figure 16.16.   Program to analyze dental measurements data as multivariate repeated measures

```
proc glm data=wide;
  class sex;
  model mm1 - mm4 = sex/nouni;
  repeated age 4;
run;
```

The REPEATED statement in Figure 16.16 groups the four dependent variables, MM1-MM4, together and defines them as four different *values* of a repeated measures *variable* called AGE. Output from this program, as in Rao's Example 16.8, is shown in Figures 16.17-16.20.

Figure 16.17 shows Class Levels Information for both the between- and within-subjects variables. Figure 16.18 gives the test for the effect of the between-subjects variable, SEX. Comparing this to the bottom line in Figure 16.12, we see that both approaches give the same results: the sex difference is significant at $p = 0.0054$.

Figure 16.17. Multivariate repeated measures analysis: class levels information

```
                 General Linear Models Procedure
                    Class Level Information

              Class      Levels      Values

              SEX           2        boy girl


         Number of observations in data set = 27


         Repeated Measures Analysis of Variance
            Repeated Measures Level Information

   Dependent Variable      MM1      MM2      MM3      MM4

        Level of AGE         1        2        3        4
```

Figure 16.18. Test of between-subjects effects

```
              Repeated Measures Analysis of Variance
           Tests of Hypotheses for Between Subjects Effects
```

Source	DF	Type III SS	F Value	Pr > F
SEX	1	140.46485690	9.29	0.0054
Error	25	377.91477273		

Figures 16.19 and 16.20 give alternative tests of the within-subjects variable main effect and the interaction involving the within-subjects variable. Four commonly used multivariate statistics, Wilks' Lambda, Pillai's Trace, Hotelling-Lawley Trace, and Roy's Greatest Root, are computed for both the main effect and the interaction, as are corresponding F-tests for each statistic. Explanation of these statistics can be found in textbooks on multivariate statistical methods. Their interpretation is clear in this context: the AGE × SEX interaction is marginally significant ($p = 0.0696$) and there is a pronounced AGE effect ($p = 0.0001$). Comparing these F-values to the corresponding ones in Figure 16.12, we see that they are quite close, indicating that any departures from sphericity in the data is slight.

Figure 16.19. Multivariate analysis of variance for repeated measures

```
         Manova Test Criteria and Exact F Statistics for
                 the Hypothesis of no AGE Effect
     H = Type III SS&CP Matrix for AGE    E = Error SS&CP Matrix

                 S=1      M=0.5     N=10.5

Statistic                  Value        F     Num DF   Den DF   Pr > F

Wilks' Lambda            0.19479424  31.6911      3       23    0.0001
Pillai's Trace           0.80520576  31.6911      3       23    0.0001
Hotelling-Lawley Trace   4.13362211  31.6911      3       23    0.0001
Roy's Greatest Root      4.13362211  31.6911      3       23    0.0001

         Manova Test Criteria and Exact F Statistics for
               the Hypothesis of no AGE*SEX Effect
   H = Type III SS&CP Matrix for AGE*SEX   E = Error SS&CP Matrix

                 S=1      M=0.5     N=10.5

Statistic                  Value        F     Num DF   Den DF   Pr > F

Wilks' Lambda            0.73988739   2.69527     3       23    0.0696
Pillai's Trace           0.26011261   2.69527     3       23    0.0696
Hotelling-Lawley Trace   0.35155702   2.69527     3       23    0.0696
Roy's Greatest Root      0.35155702   2.69527     3       23    0.0696
```

Figure 16.20. Adjusted tests of within-subjects effects

```
              Repeated Measures Analysis of Variance
           Univariate Tests of Hypotheses for Within Subject Effects

Source: AGE
                                                     Adj  Pr > F
    DF   Type III SS  Mean Square   F Value  Pr > F   G - G    H - F
     3   209.4369739  69.8123246     35.35   0.0001  0.0001   0.0001

Source: AGE*SEX
                                                     Adj  Pr > F
    DF   Type III SS  Mean Square   F Value  Pr > F   G - G    H - F
     3    13.9925295   4.6641765      2.36   0.0781  0.0878   0.0781

Source: Error(AGE)

    DF   Type III SS  Mean Square
    75   148.1278409   1.9750379

                 Greenhouse-Geisser Epsilon = 0.8672
                 Huynh-Feldt Epsilon = 1.0156
```

Finally, Figure 16.20 shows the univariate tests of the within-subjects variable and the interaction involving it, which are the same as in

Figure 16.12, plus the Greenhouse-Geisser adjustment factor and the Huynh-Feldt adjustment factor and the *p*-values associated with the two versions of the adjusted *F*-tests. In this case, presence of a very slight departure from sphericity results in the *p*-value for the *F*-test for the AGE × SEX interaction using the Greenhouse-Geisser adjustment being slightly higher than the *p*-value for the unadjusted *F*-value. The conclusion remains that there may be some interaction present, and that there surely is an effect of age. While in this example the same conclusions were reached using the unadjusted *F*-tests, the adjusted *F*-tests, and the multivariate *F*-tests, in other instances the correlation structure among sets of repeated measurements may be such that the three approaches would not all lead to the same conclusion. At least, the data analyst would want to obtain multivariate and adjusted univariate tests in order to compare them to the unadjusted versions.

16.5 Other Repeated Measures Designs

The dental measurements data analyzed in the last two sections was a two-factor study, one of which was a between-subjects variable and the other a within-subjects factor. Repeated measures studies may be single- or multifactor studies. In multifactor studies, there may be one or more between-subjects factors and one or more within-subjects factors. In this section, we present general approaches for analyzing any number of between- and within-subjects factors using SAS.

A single within-subjects factor

First, consider the simplest repeated measures design: a single within-subjects factor. For example, suppose that three drugs to treat a rare disease are to be compared. Since the disease is rare, it is very difficult to find very many patients on whom to test the drugs. Instead of dividing the few patients into three groups, each of which is given a different drug, each patient is given all three drugs, in random order, with sufficient time between different drugs to eliminate any carryover effects. Another example of a single within-subjects factor might be a study in which patients are observed for several time periods prior to receiving a therapy and for several more time periods during or after the therapy. Changes in the patients over time can be attributed to the effects of the therapy. Since each patient is measured at each time period, we have a repeated measures design.

It is convenient to think of a single within-subjects factor repeated measures design as a randomized complete block design, in which the subjects serve as blocks, and the treatments are applied at random to experimental units within the blocks. In the examples above, the subjects are patients and the experimental units are time periods in which observations are made. In the first example, treatments are randomly assigned to time periods, but in the second example, observations are made in time order, which is important, so the analogy to randomized complete blocks breaks down. In cases such as this, it will be important to examine the adjusted F-tests.

To analyze a single within-subjects factor repeated measures design from the univariate approach, read the data in the "long" format, identifying the subject, the value of the dependent variable, and the levels of the repeated measures factor, and write

```
proc glm;
      class factor subject;
      model y = factor subject;
```

As with the unreplicated randomized complete block design, the factor × subject interaction is left out of the model statement to serve as an error term. To use the multivariate approach, read the data in the "wide" format, identifying the r repeated measures as r different dependent variables, say v1, v2, ..., vr, and write

```
proc glm;
      model v1 - vr = /nouni;
      repeated factor r;
```

In this approach, since there are no between-subjects variables, no CLASS statement is used and no factors are listed on the right-hand side of the MODEL statement.

Sections 16.3 and 16.4 showed how to handle one within-subjects variable and one between-subjects variable. If B is the between-subjects variable and W is the within-subjects variable in a two-way factorial experiment, then for the univariate approach write

```
proc glm;
      class B W subject;
      model y = B W B*W subject(B);
      test h = B e = subject(B);
```

and for the multivariate approach,

```
proc glm;
      class B;
      model w1 - wr = B/nouni;
      repeated W r;
```

Extension of these techniques to one within-subjects factor and any number of between-subjects factors is relatively straightforward. For example, suppose in a three-factor factorial factors A and B are two between-subjects factors, and factor C is a within-subjects factor. To use the univariate approach, recall that subjects will be crossed with levels of factor C but nested within levels of factors A and B, and write

```
proc glm;
      class A B C subject;
      model y = A B C A*B A*C B*C A*B*C
                  subject(A*B);
      test h = A B A*B e=subject(A*B);
```

The

```
                  subject(A*B)*C
```

interaction is left out of the model statement to serve as an error term.

To use the multivariate approach to analyze one within-subjects and two between-subjects variables, write the observations on the different levels of the within-subjects variable as separate dependent variables on the left-hand side of the model statement, and then write the between-subjects variables on the right-hand side. For between-subjects variables A and B and within-subjects factor C in a three-way factorial,

```
proc glm;
      class A B;
      model C1 - Cr = A B A*B/nouni;
      repeated C r;
```

Multiple within-subjects factors

If *only* within-subjects factors are involved, this means that subjects are crossed with levels of all factors. To use the univariate approach, model subjects as if they were another factor, and test each of the factors using the interaction of that factor with subjects as an error term.

For example, suppose that all students in a certain Master's program take the same three courses their first semester. Each student is asked to record the number of hours per week spent on each of the three courses for each of the 15 weeks in the semester. Each of the three levels of the factor COURSE is crossed with each of the 15 levels of the factor WEEK. Think of STUDENT as if it were another factor, crossed with both COURSE and WEEK, and write

```
proc glm;
      class course week student:
      model hours = course week student course*week
                    course*student week*student;
      test h = course e = course*student;
      test h = week e = week*student;
```

Leaving the

```
course*week*student
```

interaction out of the MODEL statement causes it to be used as the error term for testing the COURSE*WEEK interaction. There is no test for SUBJECTS, but you wouldn't want to do that, anyway.

Analyzing these data using the REPEATED statement in GLM is a bit tricky. First, because there are no between-subjects factors, the right side of the MODEL statement will be blank. Then, each of the $3 \times 15 = 45$ recordings made by each student is defined as a separate variable, say $Y1$-$Y45$, and write

```
proc glm;
        model y1 - y45 = /nouni;
```

How to write the REPEATED statement depends on how $Y1$-$Y45$ were defined. If $Y1$-$Y15$ are the 15 weeks' of measurements made for Course 1, $Y16$-$Y30$ are measurements on Course 2, and $Y31$-$Y45$ are for Course 3, then write

```
        repeated course 3, week 15;
```

But if $Y1$-$Y3$ are readings for Courses 1, 2, and 3, respectively in Week 1, and so on, then write

```
        repeated week 15, course 3;
```

Finally, consider two or more within-subjects variables and one or more between-subjects variables. For example, suppose that factors W and X are two within-subjects variables and that B is a between-subjects variable in a 3-way factorial. Recalling that this means that subjects are crossed with levels of factors W and X but nested within levels of factor B, write

```
proc glm;
        class X W B subject:
        model y = X W B X*W X*B W*B X*W*B subject(B);
        test h = B e = subject(B);
```

for the univariate approach or

```
proc glm;
      class B;
      model Y1 - Yxw = B/nouni;
      repeated X x, W w;
```

where w is the number of levels of the factor W and x is the number of levels in factor X.

References

Cody, Ronald P. and Smith, Jeffrey K., *Applied Statistics and the SAS®
Programming Language, Third Edition*, NY: Elsevier Science
Publishing Co., Inc., 1991.

Delwiche, Lora D. and Slaughter, Susan J., *The Little SAS Book: A Primer*,
Cary, NC: SAS Institute, Inc., 1995.

Herzberg, Paul A., *How SAS Works: A Comprehensive Introduction to the
SAS System, Second Edition*, NY: Springer-Verlag (Captus Press),
1990

O'Brien, Ralph G., "Power analysis for linear models," *Proceedings of the
Eleventh Annual SAS® User's Group International Conference*,
1986, pp. 915-922.

SAS Institute Inc., *SAS® Language and Procedures: Usage, Version 6, First
Edition*, Cary, NC: SAS Institute Inc., 1989

SAS Institute Inc., *SAS® Language: Reference, Version 6, First Edition*,
Cary, NC: SAS Institute, Inc., 1990.

SAS Institute, Inc., *SAS® Procedures Guide, Version 6, Third Edition*, Cary,
NC: SAS Institute, Inc., 1990.

SAS Institute Inc., *SAS/STAT® User's Guide, Version 6, Fourth Edition,
Volume 1*, Cary, NC: SAS Institute Inc., 1989.

SAS Institute Inc., *SAS/STAT® User's Guide, Version 6, Fourth Edition,
Volume 2*, Cary, NC: SAS Institute Inc., 1989. 846 pp.

Index